[英国]蒂莫西·沃克 著　张雪 译

# 植物

牛津通识读本·

# Plants

A Very Short Introduction

译林出版社

图书在版编目（CIP）数据

植物 /（英）蒂莫西・沃克（Timothy Walker）著；张雪译．—南京：译林出版社，2021.1（2021.10 重印）
（牛津通识读本）
书名原文：Plants : A Very Short Introduction
ISBN 978-7-5447-8437-5

I.①植… II.①蒂… ②张… III.①植物 IV. ①Q94

中国版本图书馆 CIP 数据核字（2020）第 208113 号



**植物 ［英国］蒂莫西・沃克 / 著　张　雪 / 译**

责任编辑　许　丹
装帧设计　景秋萍
校　　对　王　敏
责任印制　董　虎

原文出版　Oxford University Press, 2012
出版发行　译林出版社
地　　址　南京市湖南路 1 号 A 楼
邮　　箱　yilin@yilin.com
网　　址　www.yilin.com
市场热线　025-86633278
排　　版　南京展望文化发展有限公司
印　　刷　江苏扬中印刷有限公司
开　　本　635 毫米 ×889 毫米　1/16
印　　张　16.75
插　　页　4
版　　次　2021 年 1 月第 1 版
印　　次　2021 年 10 月第 2 次印刷
书　　号　ISBN 978-7-5447-8437-5
定　　价　39.00 元

# 序 言

史 军

2017年5月，一段名为《海带不是植物》的短片在网络上疯传。一则科普短片之所以能“出圈”，是因为这个说法过于反常识了。不论是外形长相，还是吸收太阳光的能力，海带在我们看来都是活脱脱的植物，更不用说在生物课本上明明白白地写着，藻类植物是植物家族的重要组成部分。但如今，海带已经被踢出了植物家族，那植物究竟是什么呢？

2018年，我有幸参与小学科学课教材的编写工作。在一年级的科学课标准中有一个教学目标，让学生明白“什么是植物”。如何能讲清楚这个问题，我一时间陷入迷茫。

你可能会想，这问题有什么难的？扭头朝窗外一看，目力所及之处都有植物，挺拔的雪松，娇艳的月季，拥有灿烂金黄叶片的银杏，还有马路牙子上的狗尾草，这些不都是植物吗？确实，人类是一种被植物包围的物种，或者说我们的生活就建立在植物世界的基础之上。但是如何定义植物，如何认知植物，并不是一个简单的问题。

你可能会说，“这还算问题吗？植物就是那些需要晒太阳进行光合作用的生命啊”。可是这个世界上确实还有很多不依赖太阳光生存的植物，比如天麻就是一种像蘑菇一样生存的腐生植物；鼎鼎大名的大王花寄生在其他植物的根上，整个身体都“浓缩”成一个花朵。

你可能会辩解说，“植物就是长在土里一动不动的那些生命啊”。显然，很多植物是可以运动的。如果说捕蝇草那“捕鼠夹”式的运动不算典型运动，那跳舞草的小叶片在动听的乐曲中摇摆起来，必然要算运动了。至于是不是长在土壤里，就更不关键了。在南美洲，很多空气凤梨已经在悬空的电线上找到了自己生存的家园。

相对于理解动物，理解植物要困难得多，那是因为我们人类是动物，而不是植物。很多动物的形态特征，都能从我们自身找到可以类比的例子。但是，类比植物就很难了，毕竟我们不靠叶片吸收太阳光，不靠根系来喝水，更不靠花朵来繁衍后代。

那么，究竟什么样的生命形态才是植物呢？

我们需要了解生命的底层构架和运行逻辑。《植物》这本书就是从这些最基本的概念开始，为我们逐步展现植物这种特殊的生命形态，并以生物演化为脉络展现了不同植物类群在地球上的精彩表演——应对水体生活的光合生物，第一次尝试登陆的苔藓植物，用维管束系统征服陆地的蕨类植物，带着花粉走向干旱区域的裸子植物，以及让今天的地球生物圈异彩纷呈的开花植物。每一个适应特征的出现，都是生命演化历程中的飞跃。

虽然在本书中，作者仍然把藻类归在植物大门类当中，但这并不妨碍他为我们讲述藻类植物光合作用的特点，毕竟大多数藻类植物无论是细胞结构，还是光合作用的色素，都与真正的植物有所差别。

但说到底，植物也好，藻类也罢，其实都是我们人类在认知世界的过程中，为了方便而划定的单元。地理学上的五大洲四大洋，天文学上的星座，地质学上的地质年代划分莫不如此。从最初的人为二分法，到今天追寻自然分类的原则，人类对于生命的了解每前进一步，我们对植物的定义都会有新的理解。也许有一天，我们忽然发现，人类与植物也仅仅是殊途同归的生命表现形式而已，大家不过都是遗传信息的载体。

当然，读懂植物不仅可以为我们更好地开发自然资源提供基础，还可以帮助我们更好地与这些生命和谐相处，并且从它们身上汲取精神能量。每一天我们在上班和上学的路上都会与无数植物相遇，你可以把它们当作擦肩而过的陌生人，也可以把它们当作向其吐露心声的挚友。当然这要建立在理解这些生命的基础之上——叶片为何伸展，花朵为谁绽放，种子如何去远行，植物自有自己的精妙方案。读懂植物的智慧，也是读懂自然给我们的生存哲理。

毫不夸张地说，我们在追寻植物是什么的答案时，最终要回答的问题就是，我们是谁。当我们获取这个答案的时候，那必将是一个新纪元的开端。

而读懂这本书，就是一个很好的起点。

# 目录

第一章

# 什么是植物?

植物和爱情一样,很容易识别,却很难去定义。在英国,人们在许多自然风光优美的景区入口,都可以看到一块要求游客避免"毁坏树木和植物"的标志牌。我们肯定要问上一句,树怎么就不能算作植物。植物通常被简单地解释为一种绿色的、不能移动的有机体,它们能够通过光合作用来养活自己(即自养生物)。这是关于植物的一种启发式定义,我们可以增添更多的描述来对其进行完善。有时,植物被描述为具有以下多种特征的有机体:

1) 含有叶绿体,具有对水和二氧化碳进行光合作用并产生糖的能力;

2) 具有纤维素构成的坚硬的细胞壁;

3) 可将能量存储为碳水化合物或者淀粉;

4) 进行分裂和分化的特殊组织(即分生组织)可持续生长;

5) 细胞中含有相对较大且充满汁液的液泡。

所以,树木显然是植物,而且就算其他有机体缺少上述一种或几种特征，我们也不难将其定义为植物。例如，兰科珊瑚兰属植物春珊瑚根（*Corallorhiza wisteriana*）能开出兰花般的花朵，能结出典型兰科植物的微小种子，也具有大多数陆生植物都有的维管组织。但是，这种植物却没有绿色的叶片，因为这种珊瑚兰属于菌根营养植物，它依靠能从森林地被物的腐烂物质中获取能量的真菌而生存。正是与真菌的亲密关系令这种珊瑚兰能够实现这一生存模式，而这也是兰科植物在不同程度上共同具有的特征。与之类似的是生长在牛津地区查威尔河畔的欧洲齿鳞草（*Lathraea clandestine*），它的花朵让人联想起毛地黄，但它既没有嫩枝也没有叶片。它的花朵直接从土壤中冒出来，因为这种植物的根能够渗入柳树的根，并从柳树的维管组织中转移营养物质。这两种植物都失去了光合作用的能力，但它们仍然属于植物，因为它们和光合作用植物有许多其他共同特征。

上述植物定义所存在的问题就是过于狭隘，因为它并没有将部分生活在水中的藻类考虑在内。为了给植物一个合理、明确的定义，我们需要考虑生物有机体的分类方式。在生物学中，相似的个体被归为同一个物种，而相似的物种又被归为同一个属，以此类推进而归类为科、目、纲、门、界。这种分类等级中的每一层次都可被称为一个分类单元，而关于分类的研究就是分类学。在19世纪之前，分类学家们尝试创建一种能够揭示造物主计划的**自然**分类。而自19世纪起，生物学家们就开始质疑物种是否能通过保留这些变化并将其传递给后代而发生改变和进化。

目前，生物学家们已经进行了大量的工作来构建“生命之

图1　列当属植物*Orobanche flava*是一种寄生植物。寄生植物不能进行光合作用,但可以从其他植物中获取营养

树”(或者说是系统发生),以此来展示所有生物体之间的关系。1859年,达尔文发表《物种起源》之后,这项工作便实现了快速启动,而且至今仍在进行。进化树是《物种起源》中的唯一一幅插图,而第一版图书的第13章也对分类学进行了清晰的介绍。达尔文在书中探讨了构建一种**自然**分类的可能性,但是如今“自然”意味着进化过程的揭示而非上帝的旨意。目前,分类学的基础在于达尔文所谓的共同祖先学说。一个分类单元中的所有物种只有一个共同的祖先,而且它们就是该祖先的所有后代。如果能够满足这种分类标准,那么该类群便是单系的。从物种到界,单系类群出现在分类的每个层级中。

2

如果我们将38亿年来的物种进化看作一棵分叉树,那么植物就是生命之树上的一组分支,而这组分支都能与同一个分叉点发生联系。当你试图确定究竟是哪个分叉点成为植物的起源时,争论便就此产生了。此时值得一提的是,真菌绝对**不是**植物。在生命之树中,真菌实际上是与动物并列的分支。尽管如此,在高校的院系组成中,真菌学家确实比较容易和植物学家(而不是动物学家)归为一类。

## 植物的起源

无论如何定义,植物的核心无疑是光合作用能力。但不幸的是,有些能够进行光合作用的生物体却并不被看作植物。光合蓝藻就是其中的一员。

目前,人们认为生命只在38亿年前进化过一次。当时,地球作为生物环境与现在截然不同。没有保护性的臭氧层,也就无法吸收来自太阳的有害紫外线。此外,大气中还含有大量的二

氧化碳，而氧气成分却很少。 4

与我们如今所看到的大多数植物相比，最早出现的生物非常简单。首先，它们是单细胞生物，即原核生物。现在仍有许多原核生物存在于古生菌和细菌这两大类生物中。（另一大类生物就是真核生物，即植物、动物和真菌。）人类在距今有近35亿年历史的岩石中发现了原核生物化石。这些早期细菌的化石结构看起来与如今在世界各地多处都能见到的叠层石类似。

叠层石是一种垫形的岩石，见于温暖的浅湖边缘，最常见于咸水湖旁。这种岩石就是（仅仅）由微生物层积堆叠而成的。单细胞的蓝藻菌群漂浮在水上形成了一层黏液膜。碳酸钙沉积在黏液上，蓝藻随之迁移到岩石表面，继而形成一层新的黏液膜。这些交替的岩层和黏液膜逐渐演变成化石，细菌也就被包裹在岩石中。所以很显然，原核生物可能早在38亿年前就已经完成了进化，但想要确定这些早期生物如何获取赖以生存的能量却并不容易。有些生物可能合成了酶来分解矿物质，但这种能量来得太慢。目前已有有力证据表明，这些叠层石化石中的蓝藻能够捕获太阳的能量，并以大气中丰富的二氧化碳为原料来合成有机碳。这一证据基于这样一种事实，即相较于大气中存在的$^{13}C$，负责从二氧化碳中捕获碳元素的酶会优先结合碳的另一种同位素$^{12}C$。因此，如果碳化合物中含有这两种同位素的比例与大气中不同，那么这些化合物就是光合作用的产物。在格陵兰岛的岩石中所发现的碳化合物就具有光合作用所产生的 5 碳同位素比率。

我们所熟悉的光合生物通常利用水来产生电子，水中的氧气随后以气体的形式释放到大气中。人们认为最早的光合蓝藻可能采用硫化氢（$H_2S$）而非水（$H_2O$）来作为光合原料。根据目前的推断，蓝藻在22亿年前产生了大量的氧气并积累在大气中。这件事看起来不起眼，但蓝藻开始利用水作为电子供应这一事实最终导致了大气中氧气水平的上升，从而使得有氧呼吸以及大多数生命的存在成为可能。氧气的产生还有另一重影响，即上层大气中臭氧层的形成。现在臭氧层的缺失已经引起了人们的注意，它的保护作用在生物学上至关重要。在臭氧层出现之前，叠层石中的黏液层可能有助于保护蓝藻，水生环境也能为其提供部分保护作用。

回顾前文，早在20亿年前已有大量的原核生物蓝藻可以通过光合作用产生氧气，但是仍然没有可以称之为植物的生物出现。植物的进化还缺少一件必定发生过我们却并不知晓全过程的事件，那就是第一个真核细胞的形成。真核细胞比原核细胞具有更完善的内部组织，它们具有被膜包裹的细胞器，例如细胞核和线粒体，以及植物所特有的叶绿体。细胞器就是细胞内的小型“器官”，它们在细胞内各自承担着特殊的功能。

科学家们认为，在27亿年前，某种不明单细胞原核生物吞噬了另一种原核生物，但并未将其分解。被吞噬的细胞仍然保留着自身的细胞膜，并将部分（并非全部）基因并入宿主细胞的细胞核中。这种吞噬衍生而来的“合作”关系被称为胞内共生（endsymbiosis）。这种早期真核生物（即原真核细胞）依赖于代谢独立生活的蓝藻的光合产物而生存。胞内共生学说的证据十分简单：细胞器有两层细胞膜——一层属于自己，另一层则属于将其吞噬的宿主细胞。最早发生胞内共生现象的时间推断依据也很简单：所有真核生物的独特特征之一就是能产生甾醇。在真核生物死亡并被分解后，甾醇便被转化为甾烷，并在岩石中存留很长时间。拥有27亿年历史的岩石中含有甾烷，所以其中也有少量的死亡真核生物，但是并没有完整生物体的化石。

多年之后，真核生物的多样性增加导致进化谱系产生了许多其他物种（包括现存物种和灭绝物种），但没有植物。然而，在吸收了一种原核生物之后，原真核细胞又吸收了另一种原核生物，而这次的原核生物是一种光合蓝藻细菌。和之前一样，被吞噬的生物体成了一种细胞器，而部分（并非全部）基因也并入了宿主细胞的细胞核。和之前一样，这种细胞器也具有双层膜结构，它就是我们如今所知的叶绿体。



疑似真核生物结构的最古老的化石证据存在于一块21亿年前的岩石中。这种名为卷曲藻（*Grypania spiralis*）的生物已没有现存的后代。它看起来有点像是一种藻类，所以人们认为（或是希望）它可以进行光合作用。这种生物的直径有2毫米，其体积大到足以成为现今部分藻类的祖先，但我们无法证明它就是这些藻类的祖先。就现存分类单元中的光合真核生物而言，第一块无可争议的化石发现于12亿年前的岩石中。这种名为*Bangiomorpha pubescens*的生物是一种红藻，并且因其与现存的红藻头发菜（*Bangia atropurpurea*）相似而得名。除了外形类似之外，这两种红藻还具有相同的栖息地——陆地边缘和水域。

7

*Bangiomorpha*的重要性还有另一层原因：作为目前已知最古老的多细胞真核生物，其细胞不仅具有特定功能，而且其中一

种功能就是进行有性生殖。在生命之树的不同分支中，多细胞化是不止发生过一次进化的重要生物学事件之一。动物与植物距今最近的共同祖先是单细胞生物，但现在这两类生物中都是由多细胞生物占主要地位。

*Bangiomorpha*化石结构的良好程度使得人们有可能重现它的生命周期，而且它与部分红藻的生命周期非常类似。在这类植物中，孢子萌发并生长为多细胞体。孢子只含有一组染色体（也就是说孢子是单倍体），所以藻类植物是单倍体。这种植物的基部具有固着器，可以令植物体紧紧附着在岩石上。植物体的顶端则变得扁平，这样它在向上生长的过程中就能捕捉到更多阳光。原植体中的部分细胞分化为单倍体配子，这便是有性生殖的先决条件之一。



大约在21亿年前到12亿年前之间，最早的光合真核生物出现了。这些生物也是最早的一批植物，生命之树这一分支上后来出现的每一种生物都是植物。定义了植物分支的这两次胞内共生事件只发生过一次，也被看作最早的胞内共生现象。人们已经从DNA序列的分析中获得了证据。这种技术自1993年来已成为解决此前各种复杂的进化问题的重要手段。

正如本书所述，植物是单系类群。有趣的是，人们越来越明确地发现，还发生了一种次生的胞内共生，其中有些真正的植物被并入非植物生物，由此而产生的生物体也不属于植物。其中



我们最熟悉的或许就是海带等褐藻。所以说，在海滩观察潮水退去后留下的“海草”，其中红红绿绿的是植物，而褐色的那些则是动物。

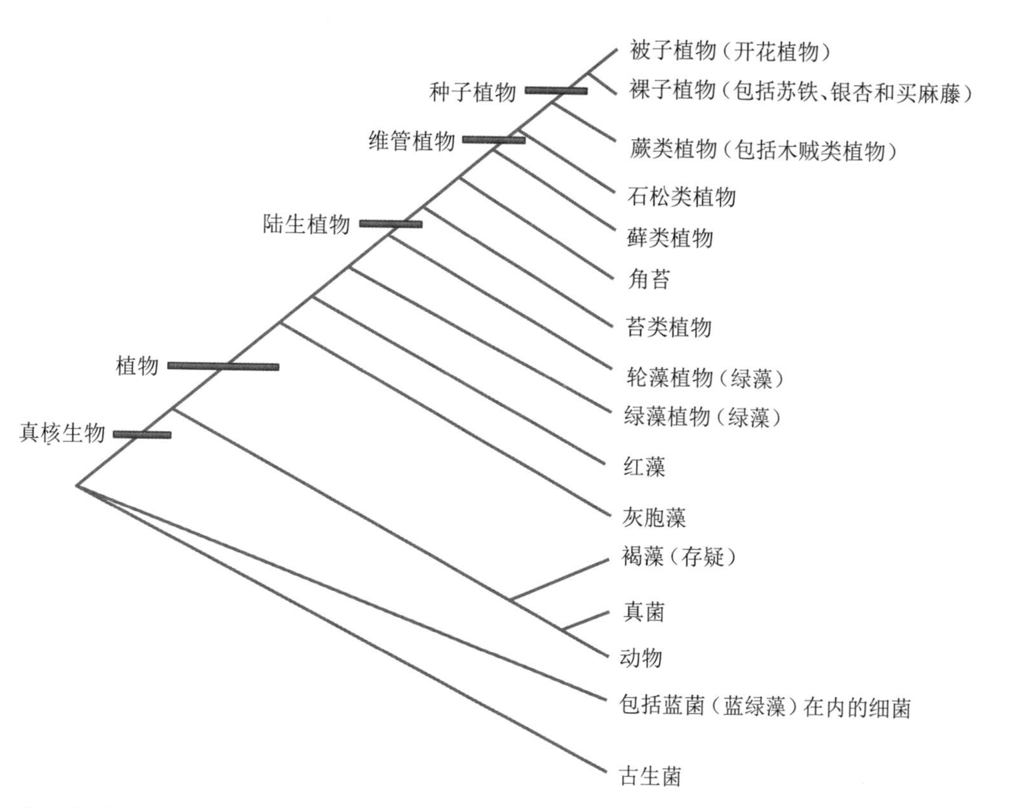
被子植物（开花植物）
裸子植物（包括苏铁、银杏和买麻藤）
蕨类植物（包括木贼类植物）
石松类植物
藓类植物
角苔
苔类植物
轮藻植物（绿藻）
绿藻植物（绿藻）
红藻
灰胞藻
褐藻（存疑）
真菌
动物
包括蓝菌（蓝绿藻）在内的细菌
古生菌
种子植物
维管植物
陆生植物
植物
真核生物

图2　植物生命进化树

## 现存最古老的植物

在生命之树中，植物分支里最低的侧支由非常古老（至少12亿年）、非常稀少（13个物种）、非常微小（显微级别）的淡水藻类组成：灰胞藻属、蓝载藻属和灰粘毛藻属——这三个属统称为灰胞藻。在系统发生树上，最低分支中的生物与其他生物相比发生了最低程度的改变。这并不意味着它们看起来与最早的祖先完全一样。人们已经具有证明其基础地位的双重证据。首先，除了包裹叶绿体的双层膜外，灰胞藻的叶绿体还具有肽聚糖层。它与细菌的被膜类似，为了与其他植物的叶绿体相区别，人们通常将灰胞藻的叶绿体称为蓝色小体。这种肽聚糖层仅存在于灰胞藻中，由此推断它在植物进化的早期就消失了。其次，灰胞藻的质体中存在一种称为藻胆素的色素。质体是植物用于生产或储存重要化学物质的细胞器。藻胆素仅存在于蓝菌、灰胞藻和红藻中，而红藻正属于植物分支中的下一个侧支。在这三种生物中，藻胆素于藻胆体内相结合。除藻胆素外，它们也都含有叶绿素a。

进化植物学家对灰胞藻很感兴趣，因为这一小类植物被看作最接近于原始内共生体的现存物种。在灰胞藻中，有些物种可以移动但有些不能，有些物种具有纤维素细胞壁而另外一些没有。在这类植物中，有性生殖尚不清楚。DNA序列分析所得出的结果与形态学证据相吻合，这些微观植物显然成了植物生命进化树中最古老分支之名的有力竞争者。下一个侧支就是红藻。需要再次强调的是，"藻类"这种通俗说法可用于概括一组生物，其中有些是植物（红藻和绿藻），另外一些则不是植物（蓝绿藻和褐藻）。

## 红　藻

这一分支中的物种数量大约在5 000～6 000种之间，或许能多达10 000种，其中只有一小部分生活在淡水中。红藻和灰胞藻具有许多共同特征，这些特征在随后的进化中丢失了，所以并不存在于其他植物中。这些共同特征包括藻胆素、藻胆体以及仅含有叶绿素中的叶绿素a。这些色素，而不是叶绿素，赋予了这些植物与众不同的红色。红藻也具有其他共性特征，比如以糖原（或红藻淀粉）的形式储存能量。糖原是由多个葡萄糖聚合而成的带有支链分子的大分子物质。有些物种可以分泌碳酸钙，而碳酸钙对于珊瑚礁的形成至关重要，因此这些物种也被称为珊瑚藻。所有的红藻都具有双层细胞壁。外层细胞壁因其可制成琼脂、用途广泛（包括可用于烹饪）而具有重要的经济价值。内层细胞壁则和大多数植物一样，部分由纤维素组成。

正如人们所料，红藻这种物种丰富的分类单元，其多样性也十分丰富，但它们的生命史都有着共同的模式。然而，这种模式非常复杂，并且与我们最熟知的哺乳动物的生命周期截然不同。这些熟知的内容会加深我们对植物繁殖的成见和假设，这种障碍也会让我们对植物生命史的理解比实际需要的困难得多。





第一种错误的假设是，**每种**自由生物都和人类一样需要两组完整的染色体。这种观点是错误的，我们已经介绍过，化石物种*Bangiomorpha*的生命大多以单倍体的形式存在，而并非具有两组染色体的二倍体。人们很容易产生疑问，二倍体或者单倍体是否具有优势或劣势呢？单倍体形式的生命存在一个问题，那就是任何有害突变都会表达出来，而生物体有可能因此死亡。所以，基因拥有两套副本似乎更好，因为有害突变的损害可以被良好的基因副本所压制，或者说基因的两种副本的结合或许要比单个更有优势。然而，这也是把双刃剑，因为这意味着二倍体细胞可以累积许多潜在的严重危害突变。但是，考虑到生命之树中的大多数分类单元都具有二倍体的进化趋势，这似乎是一种更稳定的进化策略。尽管如此，我们也无法否认单倍体并没有妨碍许多分类单元成功地存活了数亿年。这些类群包括红藻、绿藻、藓类植物、苔类植物和蕨类植物。

第二种错误的假设是，在生物体的生命早期阶段，有一类细胞（生殖细胞系）会负责形成配子，即精子和卵细胞。对哺乳动物以及许多其他动物而言，这种观点是正确的，但植物并非如此。通常来说，植物的发育以及植物细胞的分化是非常灵活的，当然不是早在胚胎阶段就决定好的。当园丁扦插枝条，或者毛地黄等二年生植物由营养生长转为开花结果时，这一结论就显而易见了。植物并没有生殖细胞系。相反，单倍体植物的生命史具有一个独特且有限的阶段，供其产生精子和/或卵细胞。由于这些植物已经是单倍体了，所以它们不需要在配子形成之前通过分化和有丝分裂将染色体的数目减半（有丝分裂是指染色体数目保持不变并产生两个相同细胞的细胞分裂）。生命史中的单倍体阶段可以产生配子，因此称其为配子体，而配子体（gametophyte）的词根“-phyte”就来自古希腊词语φυτόν（植物）。

第三种错误的假设是，二倍体植物会通过减数分裂过程产生单倍体配子（减数分裂是指染色体数目减半的细胞分裂）。

其实读完上一段介绍，这种假设已经错误得很明显了。事实上，二倍体植物会通过减数分裂产生单倍体孢子。不出所料，生命史中可以产生孢子的二倍体阶段被称为孢子体。这些单倍体孢子会生长成单倍体的配子体，这段生命史便不断循环。因此，植物的生命史由依次交替的阶段或者说世代所组成，即单倍体世代和双倍体世代交替存在。植物通常具有世代交替现象。

红藻（或它们的祖先）是最早进行有性繁殖的植物，这里值得一提，以便随后的内容更容易理解。接下来我们要介绍的是绵毛多管藻（*Polysiphonia lanosa*）的生命史。曾在英国和爱尔兰海岸的潮间带探索岩石潮水潭的人对这种红藻可能非常熟悉。这一物种生长在泡叶藻（*Ascophyllum nodosum*）的外侧，而泡叶藻是一直到欧洲西海岸和北美的东北海岸都广泛存在的一种褐藻（不是植物）。绵毛多管藻很可能是泡叶藻的一种附生植物，但有人记录称前者实际上会渗入后者，这样的话多管藻就是一种寄生植物。生长在泡叶藻表面的绵毛多管藻看起来就像是拉拉队员手中的花球。



绵毛多管藻的植物体可能是雄配子体或雌配子体，但它们看起来非常相似。雄配子体会产生雄配子，我们称之为不动精子，因为这种雄配子不具有鞭毛或尾部，无法自由游动。不动精子被释放到植物周围的水中，希望能够遇到雌配子体。雌配子体会产生（但并不释放）雌配子，雌配子会保留在一个叫果胞的结构内，而果胞由雌配子和受精丝组成。受精丝是一处细长的凸起，其功能就是捕捉路过的不动精子。一旦被捕获，不动精子就会将自己的细胞核献给雌配子体，从而形成合子这种二倍体细胞。

13
14

图3　褐藻、红藻和绿藻通常一起被冲上海滩，但只有红藻和绿藻属于植物

合子会封闭在单倍体的雌配子体内并发育成二倍体结构，也因此完全寄生于雌配子体。这种二倍体结构被称为果孢子体，由此我们也能推断出，这种疱状突起会产生二倍体的果孢子。果孢子被释放到水中，四处漂流并希望着落在合适的基质上，对于绵毛多管藻而言，这种基质当然就是泡叶藻的叶状体。二倍体的果孢子萌发并生长为二倍体的四分孢子体。而徒劳无益的是，四分孢子体与配子体非常相似，尽管前者是二倍体而后者是单倍体。当一个物种的单倍体世代和二倍体世代非常相似时，我们称这种现象为同形。当两种世代形态不同时，我们便称之为异形。有些红藻是同形的，而有些红藻是异形的。

四分孢子体成熟后，在其枝的顶端会形成四分孢子囊。四分孢子囊是单倍体四分孢子的形成之处，如此命名是因为它们是以四分体的形式形成的。也就是说，四分孢子囊呈三棱锥形排列，彼此互相依附。这些单倍体孢子是由二倍体植物生成的，所以它们正是减数分裂时染色体数目减半的结果。四分孢子被释放后，有幸碰到泡叶藻的叶状体便附着下来。这就是绵毛多管藻的完整生命史。

红藻的内部结构和它们的生命周期一样复杂多变，但我们可以拿绵毛多管藻作为例子来看。在每个枝上，罐状细胞沿着中心轴拉长排布并且首尾相连。这些细胞通过细胞分化过程中形成的纹孔所连接。与这些连接有关的是纹孔塞，后者能在某个细胞死亡后将连接密封起来。细胞内连接是多细胞体的重要组分之一。中心轴细胞周围排布着一层轴周细胞。这些细胞与每个中轴细胞的长度相当并与其对齐，从而使得这种植物的枝看起来像是一系列重复单元。轴周细胞周围可能还有一层更外

15

围的细胞，被称为皮层细胞。

## 绿 藻

绿藻属于非常多样且更加庞大的一类植物，这类植物被简单地称为绿色植物。人们目前认为，绿色植物都是由同一个祖先进化而来。它们具有许多共同特征，包括叶绿素的两种形式（叶绿素a和叶绿素b）以及由纤维素组成的细胞壁。大多数绿色植物都是陆生植物，其余的就是绿藻。

“藻类”这种通俗说法可用于概括多种生物，其中有些是植物（红藻和绿藻），另外一些则不是植物（褐藻）。“绿藻”这种说法也是一样，它可以指代并不具有相同祖先的两类不同的生物。在进化树中，绿藻有两个分支：绿藻植物和轮藻植物。这些植物的进化以及不同物种之间的关系尚未被完全了解；实际上，这还有一点神秘，因为在过去的二十年里，陆生植物比藻类植物受到了更多的关注。然而，其中一个分支的物种丰富程度相对而言要少得多，那就是包括轮藻目在内的轮藻植物。可以说，这类数目更少的植物的地位更加重要，因为它们是如今主导世界的陆生植物的姊妹群。





与其尝试对绿藻做出全面的总结陈述，我们不如详细介绍几个物种来阐释绿藻的多样性。绿藻生活在淡水和海水中。其中许多物种都是单细胞生物，但有些绿藻是丝状体，有些绿藻则会在某些阶段是单细胞生物，但随后发生细胞分化现象并形成多细胞集落。团藻（*Volvox carteri*）就是后者的典型代表之一。有些藻类会与一种叫作地衣的真菌形成共生关系。尽管没有真菌这些藻类也能生存下去，反之却并非如此，所以地衣是凭借真菌而不是藻类的身份而被人知晓的。此外，一种藻类还能与多种不同的真菌形成共生关系。（地衣并**不是**植物。）

许多人见到的第一类绿藻是衣藻（*Chlamydomonas*），因为它会生长在学校的池塘里。与团藻以及接下来即将介绍的石莼一样，衣藻这种单细胞生物也属于绿藻植物。衣藻有两条鞭毛，它是被广泛用于研究鞭毛运动的“模式生物”。成熟的衣藻是单倍体，只具有一组染色体。衣藻细胞可通过简单的有丝分裂进行无性生殖。在形成新的个体之前，衣藻细胞会失去鞭毛并聚集起来，继而以一种不协调的方式发生分裂，产生新的单细胞个体。

然而，成熟的衣藻细胞也可以分化形成配子。在某些物种中，雄配子和雌配子大小相同（同配生殖），但在另外一些物种中，雌配子要更大些（异配生殖）。两种配子在水中发生融合，形成二倍体的合子，合子被厚厚的细胞壁包裹住，起到良好的保护作用。这一接合孢子可以适应恶劣的条件，但当它感到条件适宜时，就会发生有丝分裂，每个接合孢子释放出四个新的成熟个体，这就是完整的衣藻生命史。





衣藻和我们接下来将要介绍的团藻同属于一个目。和衣藻一样，团藻也已经被研究得十分透彻，因为团藻能够以单细胞形态或者是上千个细胞组成的球形群落而存在。因此，如果你想要研究多细胞生物的进化，团藻是一种非常合适的研究对象。在植物中，团藻已经进化过许多次了，就更不用说独立于植物和动物的时候了。

所以，单细胞的团藻可以聚集起来形成群落，几千个细胞嵌在糖蛋白的胶状球体（即定形群体）外侧，而球体直径可达到1

毫米。这种群落作为集体而存在，每个细胞的鞭毛进行协调的运动，推动球体朝着光的方向移动。有时细胞质的联络索可以将细胞连接起来。团藻细胞可通过眼点感受光线，这些眼点通常分布在球体的单侧，我们可以借此区分球体的前面和背面（或者说前极和后极）。当部分细胞开始发生不对称分裂并形成大小两种细胞时，这种协同合作的细胞群落就能演变成真正的多细胞生物。分裂而来的两种新细胞无法独立存在。较小的细胞是体细胞，功能是依靠其鞭毛推动球体。而较大的细胞是生殖胞，它们积聚在球体的后极，分裂并形成子群落。这些新生成的群落最初就保留在定形群体内，鞭毛朝向球体内侧，但当母体破裂并将子体释放后，子体中的细胞会重新定向，使鞭毛朝向球体外侧。这就是完整的营养生殖过程。非常奇怪的是，在子体从母体中被释放出来之前，子体内有时就已经形成下一代子体了。



团藻的有性生殖也与衣藻不同。有些物种的群落可以同时产生精子和卵子（雌雄同体），而有些物种群落只能产生精子和卵子中的一种（雌雄异体）。（值得注意的是，这与二倍体植物的雌雄同株/异株的概念不同，后者是相对于双性二倍体植物而言的。）有性生殖开始时，群落中的部分生殖细胞**要么**开始产生被释放的精子，**要么**发育成保留在母体内的卵子。精子产生在精子包中，这种精囊会从母体中释放出去。已有部分证据表明，精子包会释放一种信息素，使得其他群落处于性活跃状态。当精子遇到卵子并成功使其受精后，二者会融合成包裹着合子的具有厚壁的孢子。这种二倍体孢子被称为减数分裂孢子，它能适应各种恶劣条件，但在适宜的环境下会发生减数分裂并释放单倍体的后代。

对于那些在世界温带地区海岸进行岩石潮水潭探索的人来说，比较熟悉的绿藻就是海莴苣，即石莼（*Ulva lactuca*）。这种看起来略显肮脏的植物可以凭借圆形的固着器固着在岩石上，也可以自由地漂移。除固着器外，植物体最多有餐盘大小，并且是仅有两层细胞厚度的叶状体，这使其非常容易损坏，所以它固着在岩石上时经常会被海浪击破。然而，当它生活在水中时，这种形态能令其漂浮在水上，所以这并不是个问题。叶状体中的每层细胞都是随机排布的，所有细胞都能发生分裂。这意味着，这种植物并不存在开花植物中所谓的分生组织。石莼的每个个体细胞之间毫无连接，在有些人看来这仅仅是一个群落。这种看法过于简化了，因为石莼实际上非常有条理，它就像多年生草本植物一样，在每年春天或者是叶状体脱落的时候能够从固着

图4　海莴苣是英格兰海滩上最常见的绿藻之一

器上重新生长起来。在实验室条件下，科学家们发现脱落的叶状体可以形成新的植物体，但这种情况不知能否在自然环境下发生。

19

与所有的性活跃植物一样，石莼的生命史也在单倍体世代和二倍体世代间发生交替。生命之树在此处的转折在于，石莼的单倍体和二倍体世代因外观相同而被称为同形。单倍体植物可产生配子，其间配子体叶状体边缘的细胞可以分裂并分化成有两个鞭毛的精子或卵子。除了卵子更大一些外，二者在形态上非常相似。雄配子和雌配子都能进行光合作用，并朝着光源游动（正趋光性）。这意味着配子可以游动到水面上。我们目前认为，鞭毛的作用不止是提供游动的动力；鞭毛不仅与性别身份有关，而且一旦发现了另一性别的配子后，鞭毛还能促进黏附。两种性别的配子在鞭毛基部都有眼点。两种配子的区别在于，雌性卵子的眼点叶绿体外膜上有5 500个颗粒，而雄配子只有4 900个颗粒。两种配子结合后，新生成的合子具有四个鞭毛并且呈负趋光性，这意味着它会游离水面，潜至水底的岩石上，在那里生长出固着器以及新的叶状体，只是这样的叶状体和固着器是双倍体。

20

发育成熟后，孢子体的双倍体叶状体会从边缘处产生单倍体孢子。这是减数分裂中染色体减半的结果。这些游动孢子（和合子一样）都具有四条鞭毛并且呈负趋光性。同样和合子一样的是，眼点外膜上共有11 300个颗粒，有些人认为这与趋光性有关。游动孢子继而会生长成一个固着器和一个雄/雌配子体。在苏格兰，海莴苣从海边采回来后会被做成汤和沙拉食用，而在日本则用于制作寿司。

就物种数量而言，轮藻植物比绿藻植物的群体小得多。在学生时代，我们可能会对其中一种轮藻植物十分熟悉。水绵（*Spirogyra*）是生活在淡水池塘里的丝状藻类。水绵通常生活在水下，但天气暖和时，快速的生长以及大量的有氧光合作用会让其在水面上长出泡沫状的黏糊糊的一片。在一年中的任何时候，我们都很容易因水绵的叶绿体呈拉紧弹簧般的分布将其鉴别出来。圆柱状细胞连接起来构成丝状体，单个丝状体可能达到数厘米长。水绵的细胞壁有两层：外层细胞壁由纤维素构成，而内层细胞壁由果胶构成。丝状体可断裂，但由于两部分短丝可分别长成新的植株，所以这本质上属于一种无性生殖。

成熟的水绵植株的每个细胞都是单倍体，所以这是一种配子体。有性生殖十分简单，可分为两种情况。在第一种情况中，两个不同的丝状体并行靠近。每个丝状体中的细胞都会生长出管子，管子顶端融合，在两个丝状体的两个细胞间形成一个接合管。雄细胞的内含物会迁移到雌细胞中，这就是核融合，由此二倍体合子形成并作为游动孢子释放出去。这个过程被称为梯形接合，因外形很像梯子而得名。另一种情况被称为侧面接合。当同一个丝状体的相邻细胞间形成接合管时便属于侧面接合。而后便和梯形接合一样，雄细胞的内含物迁移到雌细胞中，形成并释放二倍体的游动孢子。游动孢子继而发生减数分裂，形成四个单倍体细胞，再形成新的配子体丝状体形式。

本章的最后一个例子，第二种轮藻植物就是轮藻属（*Chara*）本身。这种多细胞植物生长在北半球温带地区的淡水池中。概括地说，这种植物的外形类似于其他水生植物（例如金鱼藻，见第五章）和部分陆生植物（例如木贼和牛筋草），但它与二者的亲缘关系并不紧密。轮藻具有丝状的中央茎，茎上有规律地间隔排列的节上长有轮生的枝。这种植物可能会自由地漂浮，但它也会通过根茎生长到池塘底部的淤泥中。茎顶端的细胞会发生分裂，位于上方的子细胞继续维持顶端细胞的功能。植物体的顶端细胞就是指每个尖端的细胞。位于下方的子细胞会发育成节细胞或节间细胞。轮生的枝就是从节细胞中生长出来的。枝要么短型且有限生长，要么长型且无限生长。

轮藻属于单倍配子体。这种植物会产生活动精子并将其释放到水中。雌配子并不会被释放，而是在配子体上维持其结构。就此形成的二倍体合子便可以——或许在我们看来应该——和石莼一样生长成二倍孢子体。然而，轮藻并不是人类，其合子直接发生减数分裂并产生了四个单倍体孢子。这些孢子漂走并首先发育成丝状的原丝体，继而再发育成更多成熟的单倍体配子体。



所以绿藻由两种独立的类别组成：绿藻植物和轮藻植物。它们并不是拥有同一个独特祖先的单系类群；准确地说，它们是生命之树枝干上的相邻分支。在生命之树的这一枝干上还余留着一个更大型的类群，它与轮藻植物拥有着独特的共同祖先。这个类群非常庞大，由大约40万个物种组成。我们对这个植物类群非常熟悉，因为它就是陆生植物。这些植物可以追溯到同一个独特且未得到鉴定的祖先，后者积累下许多特征的新组合，令其能够在脱离水环境的大多数时间里存活下来。这个祖先就是曾经的第一种陆生植物，没有这个祖先，就不会有我们现在知道的陆地生态系统，当然也不会有智人（*Homo sapiens*），不会有牛津大学出版社，不会有关于植物的这本通识读本。下一章讲述的便是陆生植物。

第二章

# 干燥的陆地生活

大约4.7亿年前，在绿藻生活了多年的海水中，有一种植物脱离海水并存活了不止一代。这就是地球自然史中的重大事件之一，人们通常将其称为征服陆地。然而，由于水中的许多植物已经进化出将自身附着在基岩上的方法，我们也可以轻易地（或许还更精确地）将这个事件描述为征服空气。所以，想要离开熟悉且舒适的海洋环境，植物到底面临着怎样的挑战呢？

第一，在空气中生存时，植物需要面对风造成的干燥环境。生活在海水中时，周围的水可以通过渗透作用进入到植物细胞中。可以说，生活在海水中，隔绝水要比吸收水更成问题。当第一种陆生植物被滞留在水环境之外时，它们不仅要减少水分的损失，还要找到一种吸收水分的方法来替代不可避免的损失。与吸收水分的问题一同到来的，就是保持水分的问题。如果考虑到部分早期陆生植物家族（如藓类植物）解决这一问题的方法，我们会发现它们仅仅是忍受干燥环境并在下雨时再补充水分，而非阻止干燥环境。毫无意外，在欧洲发现的藓类植物物种

24 中有10%都发现于英国，而欧洲半数的苔类植物物种都发现于爱尔兰，因为这两个国家因阴雨天气而闻名。所以第一种陆生植物有可能过着双重人格的生活方式，在湿润时吸收水分和营养，在干燥时则停止活动。

第二，也是与第一个问题相关的，就是第一批陆生植物必须找到允许空气（包括二氧化碳和氧气）进入其光合结构的方式。对于生活在水中的植物来说，这两种物质均溶解在水中且充足可用。叶片上可用于气体进入的任何开口也都是水分离开植物体的出口，而如今水分对陆生植物来说非常珍贵。

第三，同样与前面相关的问题就是如何找到植物所需的其他原料，比如氮、磷、钾、钙、铁、硫、镁、硅、氯、硼、锌、铜、钠、钼和硒。必须认识到，这些植物所移植的土地曾经是裸露的岩石。尽管可能存在沙粒、其他微粒以及较大石块的沉积物，但据我们目前所知，当时很可能还没有土壤。这是因为现今土壤的一种重要组分就是有机物质，即死亡有机体的残骸。土壤中也有其他的生物体，比如细菌、真菌和动物。在土壤出现之前，植物没有发育出复杂的根部系统从土壤中获取养分的进化压力。然而，当时的植物需要一个锚来固定自己，以免植物体从岩石上被风吹走或被水冲刷掉。藻类植物具有固着器，但这种结构似乎没有应用到干燥的陆地上。



第四，植物需要找到可以生存的新地点。尽管植物想要牢固地附着，但它们仍想探索其他区域以找到可能更好的位置并摆脱灾难。这需要单倍体孢子的保护机制的出现。结果就是，25 这种保护机制可能已经以孢粉素的形式而存在。这是在陆生植物孢子（和花粉）外部以及小球藻（*Chlorella*）等少数绿藻植物孢子壁中发现的一种非常复杂并且极其稳定的物质。对于研究孢粉素的人来说，稳定性所带来的问题就是这种物质很难显示出它的结构和化学组成成分，所以我们目前无法确定绿藻植物的孢粉素和陆生植物的孢粉素是完全相同的。因此，植物解决这个问题之前似乎已经有了预适应。预适应现象并不罕见，这可能是进化中的大型进展或多元创新的必要条件之一，否则进化就需要多种新型结构的同时出现。

在这种情况下，一些科学家面对的问题转而成了其他科学家的礼物，因为孢粉素的稳定性使得进化生物学家和生态学家可以根据土样乃至古岩中的花粉沉积来追踪植物的分布。植物出现在陆地上的一些最早证据就是，在距今4.75亿年的岩石中发现的孢子呈现出苔类植物的特征。这些孢子存在的年代要比最古老的完整植物（或者甚至是植物碎片）化石早上5 000万年。

第五个问题就是精子如何找到卵子的精妙问题。生活在水中的植物将一个配子（或者有时是两个配子）释放到周围的水中，配子会游动着寻找它的同伴。对于部分植物学家来说，这是陆生植物面临的巨大难题，但可能有些夸张了。这种论点认为，如果精子需要借助水来游动并寻找梦中的卵子，那么亲本植物就必须在潮湿的地方生存。现在仍有部分植物，例如藓类植物，需要水来促使配子的结合。简单观察下这些植物的生长环境，我们就会发现它们仅仅是被限制在**季节性**湿润的地区生存。藓类植物通常生长在墙头，这种栖息地可能与第一种陆生植物曾经生长的裸露岩石类似。尽管墙头在夏季非常干燥，但冬季并非如此，所以藓类植物配子的结合就发生在冬季。此外，说藓类



26

植物脆弱，因为它们需要借助水来完成精子的游动，这暗示了其他植物不需要水，而说任何植物都不需要水完全是无稽之谈。藓类植物对于水的临时需求并没有成为它们的致命缺点；这种植物已经在陆地上生存超过4亿年了。

如果将这五个问题综合起来，我们或许会发现，同时解决这些问题的一种简单方法就是成为一株苔类植物。园艺家们对这类微小植物中的地钱（*Marchantia polymorpha*）非常熟悉，地钱时常生长在播撒过种子的花盆顶部。这类植物像是绿色的肝状叶或叶状体（尽管苔类植物中有部分物种在简单的叶脉两侧排列有叶片状结构）。与藓类植物类似，苔类植物也可以耐受干燥环境。叶状体的上表面存在桶形的气道，为叶状体的上半层提供光合作用必需的二氧化碳以及呼吸作用必需的氧气。叶状体的下层用于储存淀粉等物质。地钱具有单细胞的假根，可生长到岩石的小裂缝中起到固定作用。这些假根不需要吸收水分和营养，因为植物体可以直接吸收这些物质。如前所述，孢子表面覆盖着抗腐蚀的孢粉素，而植物将其性行为限制在每年的潮湿时期，以避免直接将孢子释放到环境中的严重问题。所有藓类植物和苔类植物的一个共同特征就是这些植物不会长高，与大多数蕨类植物、针叶树和开花植物相比如此，实际上与在其之前出现的许多红藻和绿藻相比亦是如此。这是因为陆生植物还须面对海草无须担忧的另一个问题——重力。海水为藻类提供的浮力使其得以生长成大型结构。空气不会提供这种支撑力，藓类植物和苔类植物因而保持着小型结构，但这也并没有阻止它们存活了几亿年。藓类植物进化出一种非常简单的内部管道结构，它由导水细胞组成，这些导水细胞有时也存在于化石植物中，就比如后文中即将介绍的库克逊蕨（*Cooksonia*）。

问题来了，轮藻等植物或者当代轮藻植物的祖先究竟是如何进化成地钱或者是与地钱类似的植物呢？我们或许可以尝试寻找一种水陆两生植物的证据：它具有在水中自由漂浮的存在形式，但当其生活的池塘干涸时它也能在淤泥中存活。幸运的是，真的有这样一种苔类植物——叉钱苔（*Riccia fluitans*）。这种植物可被看作半陆生植物，或者一种缺环：它们具有部分暴露在空气中、生活在陆地上的适应能力，但没有一次性完全进化出这些能力。然而这纯属猜测，并没有化石证据的支持。不幸的是，植物征服陆地和空气的化石证据普遍比较贫乏，因为这些柔软的小型植物并不能形成良好的化石，虽然我们提到过孢子能够形成良好的化石，但也是非常微小的化石。

所以最早的陆生植物曾与苔类植物类似，并且摆脱了对水这种无所不在、包罗万象的生长介质的依赖。这可以被看作地球生命进化过程中的重要里程碑之一，因为这些植物不仅向外扩张并进化成我们现在随处可见的40万种陆生植物，它们还转而为动物、细菌和真菌提供了许多新型栖息地。一如既往，进化并未安于现状，而是在苔类植物祖先的基础上进化出许多其他的植物类群：有些植物已经灭绝，有些植物仍然存在。

## 茎

随着植物发生进化，不同的植物类群在不同的时间尺度上于生态学中占据了重要位置。数百万年的时间里，植物都不超过几厘米高。这其中有两层原因。首先，植物需要坚硬的内部骨架来使其克服重力、保持向上。其次，植物需要管道结构令



28

水和其他营养物质可以运送至不断增长的高度。此外，可将水在管道中抽上来的水泵也会有所帮助。如果想要增长高度，细胞壁和水压力只能做到这样了——最高只有几厘米。已有明确的化石证据表明，在4.25亿年前，世界上至少存在7个物种的库克逊蕨属植物（现已灭绝），这种植物茎的长度从0.03～3毫米不等。

化石显示，这些植物由二歧式分叉的茎构成，茎的顶端具有一个孢子囊，意味着这处于生命史的孢子体阶段。（不幸的是，目前没有化石证据可被认为是配子体。）在叉枝的表面可以见到使得气体进入和水分损失的气孔。如果水分充足，且蒸腾作用也可在狭窄的毛细管状的管道中拉升水分的话，水分损失并非一无是处。在库克逊蕨化石中，科学家们发现了利用这种方式吸入水分的管道。这种组织被称为木质部管，尽管这种组织或许更像是现存藓类植物中特化的导水细胞，而非现存维管植物中的木质部。人们认为，气孔造成的水分损失足以拉动水分沿着茎移动到植物体顶端需水的部位。

然而，如果植物想要长得更大，它们就得想出一些新办法，或者像时常发生的那样，翻开工具箱看看基因遗传中有没有已经存在且符合目的的事物。植物发现了木质素，这种物质最早出现在一些红藻中，而后便被弃用了。红藻究竟需要木质素来做什么仍然无人可知。我们这是将木质素出现在树木中的进化过程拟人化了，但这也对植物所采用的主要策略之一进行了说明，那就是植物为了存活下去就必须利用它们拥有的手段，因为植物并没有逃跑的选项。

为了理解木质素的重要性，我们需要先折回到植物的定义。

植物的定义特征之一就是具有细胞壁。细胞壁有诸多功能，比如限制细胞的容量以防爆炸、防止部分小分子的进入等。细胞壁的基础形式并非十分坚硬。植物的初生细胞壁由两层组成：外层的胞间层以及初生壁本身。胞间层由果胶这种多糖的复杂混合物构成。胞间层的功能之一就是将多细胞植物中的细胞维系在一起。当植物的果实和叶片脱落时，也是胞间层断开了连接。此外，它在果酱制作中也十分重要。初生壁是由纤维素和半纤维素的混合物构成的。前者可能是地球上最为普遍的非化石有机碳分子，占据了所有植物体的三分之一。纤维素由很长的葡萄糖分子直链构成。与之相比，半纤维素则由具有支链的较短分子组成，并且其组成成分不仅仅是葡萄糖。初生壁的结构强大且灵活，当里面的细胞处于完全膨压时，整体结构就会非常坚硬。灵活性也十分重要，因为初生壁要随着细胞的生长而生长。细胞壁中的纤维素微纤丝的朝向决定了细胞的生长方向。然而，当我们看到植物枯萎时，如果细胞膨压由于失水而下降，那么植物将变得非常柔软无力，细胞壁也会发生弯曲和折叠。

每次水分供应不足时就枯萎的植物永远也无法长高，除非植物生长的地方从不停止下雨。这时木质素又可以发挥作用，以组成植物的次生细胞壁。木质素十分奇怪，它的丰度仅次于纤维素；可能占到木材干重的30%，以及非化石有机碳的30%（直到所有的森林被砍光）。从化学角度来说，木质素的奇怪之处在于它缺少一致的结构。然而这并不重要，因为它能完成重要任务。木质素需要完成的任务在每个组织中都有所不同，因而它的结构也有所不同。在植物的许多部位，木质素扮演着胶

30

合剂的作用，填充了纤维素、半纤维素和果胶分子之间的空隙。木质素使得细胞壁非常坚硬。然而，木质素还可以在种子中作为储存设施，或是在含有许多木栓质时发挥防水功能。这有一个额外的好处，那就是与细胞壁的其他成分相比，木质素是疏水的，由木质素排列成的管道要比纤维素排列成的易渗漏的管道更为有效。

木质部管道的组成细胞都失去了生命组分，因而是空细胞。木质部细胞可分为两种基本设计。最早出现的（因而也被认为是更原始的类型）是管胞，继而出现的便是导管分子。与导管相比，管胞更长、更薄。管胞具有木质化的细胞壁，细胞壁中有螺旋状的增厚层。管胞通常与薄壁组织有关，后者有时也被称为基本组织。在管胞连接的地方，它们具有两个平面互相倚靠形成的长长的楔形末端。这有助于防止水流在茎内上升时连续性发生中断。管胞倾向于聚成束状来发挥功能，水从一个管胞渗向另一个，有助于防止水流中含有气泡。

与管胞相比，导管分子更短、更宽，也更难搭建。它们非常紧密地排列起来，与茎内上下细胞相连的端壁发生穿孔，使得水分通道更通畅。端壁有多种穿孔形式，但它也需要为水分运输速度的增加而付出代价，那就是水流也更容易发生中断。导管31最早发现于开花植物，也在很长一段时间内被认为是开花植物在当前大获成功的原因之一。然而，人们目前在多种植物中都发现了导管，例如蕨类、木贼类、石松类植物等，所以导管也和管胞一样发生了几次进化。考虑到世界上最高、最大的树木都具有管胞，很难认为管胞与导管毫无关联。人们认为360英尺高的北美红杉已经处于当前植物工程性能的极限。值得注意的是，我们所称树木的生长习性已经进化了许多次。我们很难对树木进行精准的定义。在法律上，树是指直径至少3英寸、距地面至少5英尺高的单茎木本植物。从植物学角度来说，树仅仅是指中间竖有一根枝条的植物——但枝条的组成可以多种多样。

不可避免的是，在这些坚硬、厚重的木质化茎的进化过程中，出现了许多其他问题，不仅包括如何直立生长、如何快速摄取水分和营养以供给植物顶端，同样重要的还有如何从植物顶端将光合作用产物运输到需要能量的根部。植物需要更好的根部，我们将在后文中对此进行讨论，但光合产物的运输也需要更多的管道。

在陆生植物曾经都是藓类植物和苔类植物时，植物中并没有大量的细胞分化。大多数细胞都能进行光合作用，所以能够自给自足。而那些不能进行光合作用的细胞距离光合细胞太近，所以只要有简单的胞间运输通道便足够了。植物一旦发生木质化，地上部分与地下部分的距离就太远了，以至于扩散作用无法生效。如果木质部是上行的运输系统，那么植物也需要下行的运输系统，这就是韧皮部。

32

韧皮部和木质部之间有一个简单的区别，那就是韧皮部细胞都是活细胞。作为一种植物组织，韧皮部由三种细胞类型组成。首先便是筛管，植物的汁液可通过这种细胞流动。筛管分子没有细胞核，但它们确实具有细胞质以及内部的少数细胞器，使其能够发挥功能。筛管像管段一样排列起来。筛管分子的端壁就像筛子，孔中可通过大于一般胞间连丝的物质。胞间连丝是将细胞彼此连接起来的小管道，而它在筛管中则连接着少量通向伴胞的细胞质，这就是韧皮部中的第二种细胞类型。伴胞具有高于平均水平的线粒体数量，因为它需要为筛管分子提供能量。伴胞中高度分化的一类被认为是一种传递细胞，它具有相当盘绕的细胞壁，使其能够从周围的细胞壁空隙中有效地聚集细胞溶质。韧皮部中的最后一种细胞类型就是以薄壁细胞、石细胞和纤维形式存在的基本组织。石细胞是一类坚硬的细胞，尤其常见于地中海环境中在夏季经历严重水压力的植物中。厚壁组织中的细胞几乎全是增厚的次生细胞壁，中央有少量的细胞质。



许多植物的生命都很短暂，它们一旦产下后代便会死去。其他植物则能够年复一年地繁衍后代。其中一些这样的多年生植物变成了木本植物，能够生存许多年。这需要植物的茎不仅能抵抗害虫和疾病，周长也要能够增长。这种支架结构无须存活，许多树木的茎干也确实不是活组织。当树木被砍倒后，我们通常可以通过计算树干中同心环的数量来衡量植物的年龄。这些环状组织是由茎尖后方最早发育成的环状维管形成层产生的。形成层是一类分生组织，可发育成其他类型的成熟组织。维管形成层的管道在内部形成木质部，在外部形成韧皮部。它还能向侧方分裂，跟随茎干周长的增长而增长。

33

在有些树木中，这种衡量方法并不奏效，因为这些树木中的维管组织并不是茎周围的连续环，而是随机分散在茎中的木质部和韧皮部的束状排列。棕榈就是其中一个例子。棕榈的芽没有如上段描述的次生侧生分生组织。生长锥具有顶端分生组织，它的后方存在生长为叶片的分生组织——叶原基。然而，还有一种结构叫作初生加厚分生组织。在新萌发的棕榈种子长出芽时，这种组织很小，但随着茎变粗，初生加厚分生组织也会变粗。它的功能就是产生维管束，就像压面机生产意大利面一样。如果年幼的棕榈树生活在最适条件中，生长锥就会越长越粗，直到达到该物种的最大直径为止。植物体继而会开始生长，并产生合适的树干。然而，如果植物在未来遭受一段略差于最适条件的时段，分生组织的宽度就会下降，导致茎缩小。然而，如果生长条件能够再次改善，那么分生组织也能扩张到先前的水平，树木因此也能长到腰部那么粗。

当种子萌发时，最早出现的结构通常是嫩芽，因为吸收水分是幼苗的第一需求。如果我们把种子看作包裹在坚硬外壳中、口袋里还装有盒饭的胚胎，那么先长出根就是个不错的策略。然而，盒饭只能在一段有限的时间内为胚胎供给，而且这个胚胎也不能回家找妈妈。它必须生产自己的食物，为了做到这一点，它就需要进行光合作用，因此也就需要叶片（或者类似叶片的组织）。



## 叶

茎尖的组织排列与根尖完全不同。例如，茎尖没有“茎 34 冠”，因为挤过空气应该不会造成损伤。这并不是说，许多植物中的生长点没有受到保护。保护生长点的简单方法就是让嫩叶及其叶柄排列在分生组织周围。茎的分生组织会发育成侧生附器（比如叶以及更多生长为侧枝的茎）和终端器官（比如花）。这与根不同，因为根通常不会产生任何分化结构，而只会产生更多的根。根中细胞发育的控制来源于根中细胞的位置，而在茎中，有更高级、更复杂的基因活动控制着组织的运转。

我们已经介绍了茎及其组成细胞的结构，但茎主要是支持系统。尽管绿色的茎也会为植物的光合作用活动做出贡献，但这远远不够。只有在仙人掌以及其他部分肉质植物中，全株植物才需要光合作用的茎，因为在水分短缺时，叶片作为一种可行策略实在是太浪费了。然而，大多数植物都会分化出叶片。每个物种的叶序通常是不变的，并且是由生长锥的相互作用所决定的。茎的分生组织处于一种断裂与补充的恒定状态。当一片分生组织从侧面分离时，它就会变成一处叶原基，继而生长出基部带芽的叶片。下一个叶片会在另一片分生组织分离后形成。而这个叶片的生长位置取决于前一处叶原基的抑制作用有多强。根据所考虑的植物物种叶序的不同，最小抑制作用的生长点会分布在茎周围的不同角度。在互生叶序的植物中，这个角度是120°；在对生叶序的植物中，这个角度是180°；在轮生叶序的植物中，这个角度大约是137°。



我们通常都会认为自己了解叶片是什么。叶是一种扁平的绿色结构。我们可能记得课堂上学过叶片的上表面细胞具有厚角质层；下表面细胞具有气孔以完成蒸腾作用，同时角质层较薄；中间则有像砖块一样排列的上层细胞（栅栏叶肉），以及更加开放的下层细胞（海绵叶肉）。在这些结构中穿过的是由木质部、韧皮部以及可能有部分木质化的硬化组织构成的叶脉。这是一种很好的广义上的叶片结构，但在这里，正如植物生物学的诸多方面一样，叶片并没有真正的默认设置，因为植物已经适应了太多不同的生态位。

事实上，叶是茎上的侧生结构，在叶从茎上长出的叶腋处，芽可能生长为营养枝或者是花序。叶是一个具有预先决定好的有限尺寸的确定结构。有些“叶”实际上是刺状突起（例如豆科

图5　太阳瓶子草（*Heliamphora*）的叶已经过改良，适用于捕捉昆虫



植物的部分成员），有些叶会发育成瓶状以捕获愚蠢的昆虫（例如太阳瓶子草），还有些叶呈卷须状用于辅助攀爬。严格来说，事情恰恰相反，刺状、瓶状、卷须状的可能是叶片。这是对同源原理的一个简单诠释：功能并不能定义结构，而结构的特性是由其相关位置和发育起源所决定的。 36

同样地，有些叶片看起来不像是叶片，也有些其他结构看起来像是叶片却不是叶片。英国植物假叶树（*Ruscus aculeatus*）的花以及果实生长在叶片的上表面。同样，马德拉群岛的攀缘植物仙蔓（*Semele*）的花生长在叶片的边缘上。看到这类事物时你必须要非常小心，因为绿色的扁平物并不都是叶片，而叶子上也不能长出花朵。这些开花的叶只不过是在外形上类似叶片的扁平的茎。

图6　假叶树，又名“屠夫的扫帚”，看起来像是花朵长在叶片上。这些“叶片”实际上是扁平化的茎

## 枝

藓类植物和蕨类植物的茎通常一分为二地开叉。这是相对容易做到的，因为它只需要顶端分生组织区域分裂成两半。在大多数种子植物中，茎分生组织在植物生长过程中保持基本相同的大小，而侧生分生组织脱落并形成带有芽的叶，而这些芽又包含着侧生分生组织。在一些植物中，叶原基和侧生分生组织一起离开茎分生组织，但结构并不相同。在其他植物中，只有叶原基与茎分生组织分离。随后在叶片开始生长和分化时，其中一些细胞才恢复成分生组织。

## 根

我们已经注意到，小型的苔类植物和藓类植物具有细小的单细胞假根，这种微型的根状结构由单个细胞构成，可使植物吸附到岩石上。在植物界，包括轮藻植物在内，假根随处可见，所以这可能也是一例陆地生活的预先适应。近期有证据表明，藓类植物小立碗藓（*Physcomitrella patens*）的假根发育调控与开花植物的根毛发育调控受到同样的基因控制。

微小的假根本身绝不可能具备为欧洲蕨等大型多年生蕨类 38 植物单独提供水分和营养的能力，更不用提仙人掌、红杉或是橡树了。根部生物学是相当受忽视的研究对象；在这种情况下，眼不见确实是“心不烦”。就像植物的地上部分在不同的物种中有所不同，植物的地下部分也是如此。大多数植物物种根部（80%～95%）的一处相同点就是与细菌或真菌的密切关系。这些丛枝菌根为植物提供营养而令其受益，特别是提供了磷酸盐。

反之，真菌也从根部的部分细胞中摄取糖分。

这种类型的关系已经出现在化石记录中，似乎是从我们认为不具有真正根的植物的地下部分进化而来。现在我们认为根至少经历了两次进化，且不包括假根。最早的根发现于3.5亿年前的化石中。这些根所属的植物隶属于石松类植物。这种类群在进化树上的分支位于藓类植物和蕨类植物之间。石松类植物是最早生长成严格意义上的树木的植物，所以它们一定具有能够供给树冠的根部系统。这些古代石松化石的根部与更为现代的石松根部不同，前者具有由木质部和韧皮部组成、周围是薄壁组织的单一中央核心，以及具有根毛的表皮。

到蕨类植物开始产生根时，植物中已进化出一条不同的发育途径。在大多数蕨类植物和种子植物中，根尖是由四个不同的区域组成的。位于顶端的根冠可以保护更加精细的结构在根系穿过土壤时免受损伤。根冠后方就是分生组织，细胞分化就在此处发生。位于分生组织中心的是静止中心，人们认为此处可以调控周围的分裂细胞的分化（与分裂相对）时间。随着顶端逐渐脱离，分生组织产生的新细胞随后会变成伸长区。这些细胞在伸长区完成扩大后会在分化区找到自己的位置，并发育成表皮、皮层、内皮层或维管组织。在更后方，多年生植物会产生侧生分生组织的管道。这种形成层将形成次生木质部（管道内部）和次生韧皮部（管道外部），因此形成更大的永久根。这些根继而会开叉并形成复杂的根部系统，就像我们看到狂风连根拔起的成熟树木那样。尽管这种类型的根部系统很常见，但还有另一种根部类型叫作不定根。这种根生长自植物的某个部位，而非自种子胚胎长出的初生根。这种情况发生在单子叶植



39

物中，也解释了成熟的棕榈树即便只有少量的根也能进行移植的原因。更多细节我们将在第五章中进行介绍。

现在我们已经了解了各种各样的陆生植物，其高度从贴地的苔类植物到高耸的北美红杉不等，体型从微小的浮萍到巨大的桉树不等。然而，无论植物多大或多小，它们的生命都有着同一个目标：在死去之前繁衍出更多的植物。在先前讲述的世代交替的许多变化中，它们就是这么做的。让我们翻开第三章吧。 40

第三章

# 植物的繁殖

生产与亲本类似的更多新个体是生物体的定义特征之一，植物也毫无例外地遵守这项原则。当植物最初迁移到陆地和空气中时，它们所具有的生殖系统只适用于水中，但这在水分并非无处不在的地方仍然能发挥功能吗？显然，答案是肯定的，但植物仍需要做出一些变化。

在我们见到的大多数（如果不是全部）陆生植物的主要类群中，繁殖更多植物有两种从根本上完全不同的方式。更简单的方法是进行营养复制。这种方法的优势在于：要比有性生殖更加简单；只需一棵植物就能进行繁殖；更加迅速；如果亲本对环境的适应极好，那么后代也将如此。我们发现营养生殖（或无性生殖）是入侵物种的普遍特征，所以这种策略显然具有强大的短期优势。植物进行自身的独立复制有许多种方式。

## 无性生殖/营养生殖

苔类植物具有一种被称为胞芽的结构。这是一种未分化的细胞团，生长在叶状体表面的胞芽杯中。当这个细胞团准备好被释放时，它便被雨滴移出并被冲刷走。当它停止移动后，便能生长成和亲本一样的植物。皱蒴藓（*Aulacomnium androgynum*）等藓类植物可由茎尖产生胞芽。这些胞芽并非无定形的团状，而是具有明显的末端，在分生组织的顶端细胞和基底细胞之间有几个细胞。这些胞芽的传播可以借助任何可利用的方式。

在英国湖区走路滑倒过的人都会对欧洲蕨（*Pteridium aquilinum*）的传播能力印象深刻。这种植物在山坡上生长的过程中，每个个体的嫩芽都与亲本植物相连接。然而，每个嫩芽都有自己的根部系统，也完全可以依靠自己的能力生存。其他一些蕨类植物可以通过叶片生产小型植物。胎生铁角蕨（*Asplenium viviparum*）和珠芽铁角蕨（*Asplenium bulbiferum*）就是两个这样的物种。

在裸子植物中，营养生殖相对罕见，但人们熟知的一个例子就是北美红杉（*Sequoia sempervirens*）。这个物种在其基部周围的环形范围内繁殖年幼的植物。当亲本植物死亡后，它就会留下沿同心环生长的年幼树木。最终，这些树木每一棵又会产生另一个环状的替代品。我们可以认为这是一种再生方式，而非植物的繁殖，因为其中没有传播迹象。塔斯马尼亚的泣松（*Lagarostrobus franklinii*）也会产生这种长寿的无性系种群，人们认为这一种群的年龄已经超过一万年，它们因此也是地球上年龄最大的生物体或基因型。

正如所有的园艺家都知道的，营养生殖在开花植物中非常常见。偃麦草（*Elytrigia repens*）、羊角芹（*Aegopodium podograria*）和田旋花（*Convolvulus arvensis*）就是三种在栽培地大获成功的杂草植物，因为它们能够由挖掘后留在土壤中的一小段植物再生

42 为成熟植物。如前所述，许多世界上最具入侵性的植物物种可以通过快速的无性生殖传播占领某个地区。这类植物就包括黑海杜鹃（*Rhododendron ponticum*）和黑莓。在前者中，植物体最低的枝条会在接触土壤时生根；在后者中，茎尖与土壤接触几周后就会生根。这种繁殖方式中有一种极其狡猾的变体是在无味假葱（*Nothoscordum inodorum*）中发现的，无味假葱在其主球茎周围以及花期中的花附近会产生小球茎或珠芽。再加上大量的种子生产，人们无法将这种植物从我们不需要的地方根除。大薸（*Pistia stratiotes*）会在侧芽末端产生完美的小植株。这些植株会因大型动物的扰乱而折断。然而，如果不受干扰，这个物种将形成巨大
43 的垫子，成为足以支撑起犀牛（虽然是小犀牛）的漂浮岛屿。

图7　假葱属植物（*Nothoscordum*）可在亲本球茎周围长出许多小珠芽，使其成为一种非常顽固的杂草

在一些植物中，有性生殖和无性生殖之间存在着一种折中的方式，叫无融合生殖，或者有时叫作孤雌生殖，这是在讨论动物时更常用的术语。在无融合生殖中，含有可用胚胎的种子在卵子没有受精的情况下产生。这意味着种子在基因上与亲本植物是相同的。对于植物学记录员来说，问题在于世界上的每个地区都可能存有该物种稍微不同的克隆体。蒲公英就是这样，具有250多种已命名的小种，但这些都只是靠营养生殖而由母体分离而成的植物。这些克隆体可能很短命，但这里对于保护有一些启发。我们是要保护每一个变种，还是应该保护维持这些小种数量的过程呢？

然而，如果栖息地发生变化，当前能够很好地适应特定的栖息地并不能保证未来的成功。虽然进化是无法预见的，但是在有性生殖过程中，通过基因突变和重组来产生变异的能力可以为进化提供修正，这也是达尔文所认为的进化核心。正如达尔文所认为的，"这是自然界的一般法则，没有哪个有机体可以世世代代自我受精"，因为这会导致使后代变得越来越弱的"近交衰退"。在这种情况下，我们来看看第二种更为复杂的繁殖方式，那就是有性生殖中精子和卵子的融合。

## 有性生殖

在第一章中，我们介绍了植物的生命史，其中涉及植物特有的一种策略——二倍体和单倍体阶段的交替，即世代交替。如果我们接受轮藻属等绿藻是与陆生植物亲缘关系最接近的物种，并且接受苔类植物是陆生植物在进化树上最低分支的现存后代，那么这两类植物之间的差异，就是植物何以迁移到陆地和 44

空气中的线索。在第二章中，我们已经讨论过植物所面临的诸多问题是如何解决的（干燥环境、支撑、寻找原料和营养），但有一个突出的问题就是：当此前在无所不在的水环境中游动的方法并不总能被采用时，配子将如何结合。

简单回顾一下，轮藻属植物可以产生许多带有两条尾巴（或者叫鞭毛）的精子，它们游向另一株植物，那里有很多卵子位于藏卵器中。精子使卵子受精（每个卵子一个精子）并产生一个合子。合子随后经过减数分裂产生四个孢子。精子、卵子和孢子都含有一组染色体（单倍体），而合子则含有两组染色体（二倍体）。这种生命史和所有的陆生植物之间有一处很大的差异，还有几处较小的差异。这种很大的差异在于，陆生植物的合子不会立即进行减数分裂；相反，它会发生有丝分裂（简单的细胞分裂）并发育成胚胎。这是一种新方式，也是陆生植物被称为胚胎植物的原因（主要是由植物学家！）。

## 藓类植物和苔类植物

所以我们从苔类植物，特别是地钱开始介绍，因为这可能是我们研究最多、理解最透彻的苔类植物。地钱是一种深绿色的叶状植物，像湿的肝脏一样闪闪发光，在潮湿的生长介质表面扩散时分叉。在叶状体表面可能有胞芽杯，每个直径约5～8毫米。然而，在一年中特定的潮湿时期，我们可以看到叶状体表面生长出伞状结构。有一些伞状结构具有完整的顶部，而有些伞状结构看起来只剩一副骨架，就好像所有的布料都被吹走了，只有辐条仍然存在。

在完整伞状结构的上表面，精子会从精子囊中游走。这些精子囊是轮藻与陆生植物之间的较小差异之一，尽管它们**看起来**与轮藻的精子产生器官没有太大区别。雄器是陆生植物中发现的两种配子囊（gametangium）之一。［英文中的后缀-angium/-angia用于描述生殖结构，因此gametangium产生gametes（配子），sporangia（孢子囊）产生spores（孢子）。］另一种配子囊是产生卵子的配子囊——颈卵器；颈卵器是轮藻与陆生植物之间的另一处较小的差异，尽管它们**看起来**和轮藻的卵子产生器官也没有太大区别。有一处显著的区别在于，颈卵器只产生一个卵子，而轮藻的藏卵器则能产生很多卵子。

所以精子利用两根鞭毛游向颈卵器中的卵子。这些颈卵器位于只有辐条的伞状结构的下方。这些植物为雌雄异株，所以每株植物只有一种伞状结构。这些植物性别的决定机制是遗传性的，并且是由异形染色体造成的：一个小的Y（雄性）染色体和一个大的X（雌性）染色体。这听起来可能很熟悉，事实上在人类中，雄性是XY而雌性是XX；但你必须记住，产生配子的苔类植物（配子体）是单倍体，所以雄性是Y，雌性是X。同时，我们再回到苔类植物的颈卵器中，卵子已经被精子受精，形成了二倍体的合子。这个合子现在分裂并成长为胚胎。它**不**会由颈卵器释放出来，而是从雌配子体中获得营养。胚胎生长并且发育成挂在柄上的一个囊；这就是苔类植物的孢子体。它不是独立生活的，而是依赖配子体度过它的“一生”。在这个囊内，孢子被产生出来并包裹在胞粉素中。这就是第一章中介绍过的坚硬外层。因此，这个囊就是孢子囊。孢子是经过减数分裂后产生的，减数分裂使染色体的数量从二倍体减少到单倍体。所有的孢子**看起来**都一样，所以我们说这种植物具有同形孢子。实际



46

上，孢子并不完全相同，因为一半孢子是雌性，一半孢子是雄性，但是它们的大小相同并且由相同的孢子囊产生。

这些孢子从破裂的孢子囊中脱落，被风吹走或是被水冲走。孢子是坚硬的小型结构，因为它们拥有遗传自水生祖先的孢粉素外壳。当孢子停靠在合适的介质上，并且温度和湿度合适时，它们就会萌发。孢子要么成长为雄性叶状体，要么成长为雌性叶状体，然后这些配子体继续长大直到成熟，当它们撑起伞状结构时，生命史便完整了。

藓类植物的生命史与之相似，但有所不同。我们已经知道藓类植物比苔类植物领先一步，因为它们具有气孔。我们可能更熟悉藓类植物，因为它们都是常见的植物，生长在城镇的墙上和花园的草坪上。生长在墙壁上的植物在我们所考虑的问题上更为有用，因为在一年中的某些时候，特别是冬末春初，这些植物会故意展示出一些类似潜望镜的小型结构——一根细细的丝状茎上支撑着几毫米长的孢蒴。这个孢蒴就是孢子囊，所以从其开口端长出的是坚硬的单倍体孢子，这些孢子看起来也都一样。我们对藓类植物配子体性别的鉴定一无所知，但在角齿藓（*Ceratodon purpureus*）中，它可能是由性染色体的遗传决定的。然而，在小立碗藓等部分藓类植物中，孢子萌发并生长为两性（或雌雄同体）配子体。

无论配子体是两性还是单性，精子囊和颈卵器都会发育，且通常位于叶状芽的顶端附近。天气潮湿时，精子从精子囊游动到颈卵器中，希望在那里找到它梦中的卵子。我们还不清楚精子寻找颈卵器的方式，但在某些物种中，颈卵器似乎会发出一种化学信号。精子使卵子受精后，合子就生长为多细胞结构，并长

47

成孢子体。在泛生墙藓（*Tortula muralis*）中，孢子体就是晚冬时节我们在墙上看到的“潜望镜”。在孢蒴中，减数分裂发生，单倍体孢子产生，生命周期就此完成。

## 蕨类植物

在生命之树上，现存植物物种的下一个主要类群是蕨类植物。如果查看蕨类植物老叶的下方，你经常能看到破裂的脓疱释放出棕色的粉末。如果你把叶子从植物上剪下来，轻轻地固定在一张白纸上，然后把白纸放在锅炉或雅家炉上过夜，第二天早上摘掉蕨类植物的叶子时，纸上就能完美地展示出蕨类植物叶片的孢子囊分布，因为棕色的粉末就是包裹在孢粉素中的上千个完全相同的孢子。这意味着破裂的脓疱就是孢子囊，而叶片就是孢子体的一部分。这显然要比生长在墙顶上的小型藓类的“潜望镜”大得多，而且在另一方面也有所不同：这种成熟蕨类的孢子体是独立生存的，而潜望镜般的藓类孢子体则完全依赖于它们下面的配子体。

所以孢子被释放并被吹走，希望能降落在一片潮湿的土壤或是一根潮湿的树枝上，或者任何湿度确实较高并且含有一些营养物质的“土壤”。孢子萌发并长成配子体。配子体是独立生存的绿色植物体，但它们不是很大。配子体在任何方向都很少超过10毫米。它们通常是扁平的叶状体，在许多情况下都与小型苔类植物没有什么区别，而且因为二者经常在相同的条件下一起生长，所以蕨类植物的配子体可能会被误认为是苔类植物。配子体可以是雄性、雌性或者两性的，这取决于物种或是环境条件。在水蕨（*Ceratopteris richardii*）中，第一个萌发的孢子 48

是两性的，可同时产生颈卵器和精子囊。然而，这些前期配子体会产生一种激素，影响附近所有新萌发孢子的发育，如果激素在最初几天（2～4天）到达这些年幼的配子体，那么配子体就完全是雄性的，只产生孢子囊，因此只有精子。这种策略的优势在于它促进了远交。另一种促进远交的方法是让配子体在不同的时间产生配子囊，即先成为雄性一段时间后再成为雌性，反之亦然。

配子体同时产生精子囊和颈卵器的一个固有问题就是它只能产生自受精卵，因此合子就是配子体的二倍体克隆。查尔斯·达尔文的生物学金句合集中有一句是“大自然憎恶重复的近亲繁殖”，正如我们将在本章中稍后提到的，大自然为避免近亲繁殖而陷入诸多麻烦——大多数时候都是如此。“大多数时候”是因为所有植物，不管是水生的还是陆生的，都面临着一个挑战，那就是它们不能移动。如果某些植物是该物种在附近地区的唯一成员，那么如果它们一定需要另一株植物的精子进行受精的话，它们就根本无法繁殖后代。所以很多植物都有一种策略：如果没有其他选择，它们就会接受自己的精子，这样至少能繁衍下一代，获得下一次远交的机会。达尔文把这种策略称为“繁殖保障”，而且这种策略非常普遍。

抛开对精子从何处游走的担忧（现在它们有许多鞭毛可以帮助移动），合子在颈卵器中形成并长成胚胎。然而，胚胎之后会在配子体上方产生一小片叶子，以及在其下方产生一小条根，而不是在某种类型的茎上长成孢子囊。这种年幼的植物最初是由配子体进行供给的，但是当它有一些叶子和根的时候，亲本配子体已经耗尽了它所有的资源，成为一副空壳——这可能也是一个对人类状况的隐喻。

到目前为止，这个故事中所有的孢子体都只产生了一种孢子；它们都具有同型孢子。人们认为，像库克逊蕨这样的化石植物也具有同型孢子。然而，现存的许多植物类群和绝大多数植物物种都具有异型孢子。孢子异型已经经过了几次进化，而第一次进化可能是在蕨类植物出现之前。在生命之树上，位于藓类植物和蕨类植物之间的是石松类植物。这种相当神秘且多样化的植物类群是最早长成树木的植物之一，也是最早产生雌雄孢子的植物。在任何情况下，我们都不能把这些孢子与配子混淆。因为是孢子，所以它们和配子一样都是单倍体，但孢子所能做的就是长成配子体。孢子不能相互融合而产生二倍体合子。



## 种子植物

陆生植物的最后一个主要类群都具有异型孢子，它们就是种子植物：裸子植物和被子植物。裸子植物包括针叶树、苏铁、银杏和一小群植物怪胎，其中包括来自纳米布沙漠的标志性奇观百岁兰（*Welwitschia mirabilis*）以及药用麻黄碱的原始来源双穗麻黄（*Ephedra distachya*）（详见第六章）。另一方面，被子植物则是开花植物。人们相信种子植物拥有一个共同的祖先，还认为它们最近的共同祖先也是具有异型孢子的，但这位祖先的形态在科研领域仍然是许多人职业生涯里的研究对象，它也激发了许多猜测。种子植物的出现可能是一个谜，但根据达尔文的说法，被子植物的起源则是一个“令人憎恶的谜”，目前仍未解开。 50

图8　日本的石松。这种植物的近亲是最早长成树木的植物之一

进化是一件复杂的事情。它没有先见之明，所以只是保留了在当前有效的机制，并没有储存很多在未来可能有用的东西。这种做法也相当冷酷。达尔文相信，没有哪种生物会为了除自身外的其他生物的益处而获得某种特性。对某种生物有利的事物可能会在未来的某个时候被另一种生物所利用，但进化不会考虑到这一点。种子就是一个很好的例子。假如种子没有在种子植物身上进化出来，便很难想象现在世界的样子。包括65亿人类在内的许多动物都把种子作为主食。然而，种子只有在给予植物优势时才能发生进化。这些优势可能是生存和传播，可能是在传播中生存下来的能力。所以种子在空间和时间上均可进行运输。但是种子是什么？它是如何产生的呢？我们已经在第二章中将种子描述为“包裹在坚硬外壳中、口袋里还装有盒饭的胚胎”。如果种子里已经有了一个胚胎，那么这个胚胎是在繁殖发生**之后**形成的。

如果你曾经在春天把车停在雪松（*Cedrus* spp.）树下，你回来时很可能会发现车上覆盖着黄色的灰尘。你甚至可能在车周围的地面上发现一些柔软的、香肠形状的球果。如果仔细看看这棵树，你会看到更多这样的小球果和一些直径几英寸的非常大的木质球果。这些木质球果内部就是这部分故事开始的地方，因为我们在其中发现了孢子囊。（事实上它被称为大孢子囊，因为它所产生的孢子要比种子植物中发现的其他类型的孢子囊产生的更大。）这些孢子并不像我们目前在每个物种中所观察到的那样被释放出来。大孢子被保留在球果中，并在此处生长成与蕨类植物配子体大小相似的雌配子体。颈卵器生长在配子体上，所以这与蕨类植物也没有太大区别，只是配子体不是独立生

52 存的，而是完全依赖于孢子体的营养。

如果这些内容还有意义的话，你应该问问自己，雄配子体在哪里？当然，雄配子体形成于其他地方，然后它们必须找到雌配子体。雄配子体生长于小孢子囊产生的小孢子中，而这些孢子囊存在于其他球果中，也就是你在汽车引擎盖上发现的那种球果。小孢子形成时带有传统的孢粉素外壳。然而，在它们被释放之前，孢子内部经历了少量的细胞分裂，另外还有两个大的气囊生长在孢粉素的外部。这整个结构就是一个未成熟的雄配子体。它未成熟是因为它还不能传播精子。这个未成熟的雄配子体希望找到一个成熟的雌配子体。但问题在于，雪松并不像蕨类植物、藓类植物和苔类植物那样，两种配子体一起处于土壤上，而是雌配子坐落在树上其他位置的木质球果里，或者最好是坐落在另一棵树上。

机灵的雄配子体通过滑翔来寻找雌配子体，并利用气囊在空中飘浮。它们希望其中一个雄配子体能降落在木质球果上，而且最好是另一棵树上的木质球果。发生这种事的可能性是百万分之一，但这已经足够了。这些能够滑翔的未成熟的雄配子体更常被称为花粉粒。当它落在木质球果上时，花粉粒从我们发挥想象为之命名的花粉滴中吸收水分。花粉粒随后长出一根花粉管，花粉管顶端就是一个精子细胞。这种精子不是由精子囊产生的；精子囊如今已成为进化的历史了。这种精子没有鞭毛，所以花粉管必须长到雌配子体颈卵器的颈部，而雌配子体一直在耐心地等待它的雄性访客。精子使卵子受精，然后产生一个合子，并长成一个胚胎；到目前为止情况还不错，只是胚胎距离合适的生长地点还很远，所以它的发育中止了，木质球果的一部分为其提供了一层厚厚的外壳，而雌配子体则承担起“旅途食物”的角色。种子从木质球果中释放出来并飘到地面上，希望能与它的母亲保持一定距离。到了地上，胚胎恢复生长并成长为一棵雪松。

53

如果这棵雪松被换作橡树、木兰或是水仙花，那么把精子和卵子聚集在一起的过程又将有所不同。这些开花植物确实能产生花粉，但它们不是位于球果中的小孢子囊，而是位于花朵雄蕊末端的花药中。开花植物的花粉类似于雪松（和其他裸子植物）的花粉，因为它也是一个未成熟的雄配子体，在寻找一个成熟的雌配子体。开花植物花粉所面临的问题在于，雌配子体及其卵子被藏在心皮内。心皮是裸子植物向开花植物进化过程中的重大创新。心皮由三部分组成：柱头（花粉的着陆点）、花柱（长度根据物种而有所不同，是柱头和子房的连接茎）和子房（包含雌配子体，每个雌配子体都有自己的卵子）。雌配子体由大孢子发育而来。这些大孢子是由子房中的大孢子囊产生的。必须指出的是，与其他陆生植物所产生的一切相比，这些雌配子体不过是配子体的一个可悲的道歉。这些被子植物的雌配子体仅由几个单倍体细胞（其中一个就是在没有颈卵器帮助下产生的卵细胞）和一个具有两个单倍体细胞核的细胞组成。我们很容易提出充分的理由来说明这种双核细胞应被授予“世界上最重要的细胞”的奖项，因为它们不仅发育成胚胎的食物，而且还成为水稻、小麦和玉米的谷粒，为全世界供应了食物。



所以花粉会降落在柱头上。花粉究竟如何精准地找到柱头可以说是植物生物学中最好的故事之一，我们将在本章的结尾处讨论这个问题。现在，我们可以假设未成熟的雄配子体已经落在柱头上，而它面临的问题才刚刚开始，因为这只是“拜见岳母”。柱头有两个问题要询问这位年轻的男性。第一，你是正确的物种吗？有些母亲没有其他人那样挑剔，或者只是在外形上有所考验，这就是为什么有时来自另一个近缘种的花粉会通过考验并有可能产生杂交后代的原因。这可能是产生新物种的一个非常重要的方式，进化可以利用或丢弃。第二个问题是，你是我吗？和所有的生物一样，近交对植物来说也是退而求其次的选择。同样，有些植物在这方面比其他植物更加勤勉。例如，十字花科的植物非常小心避免被自己的精子受精。一旦花粉恰当地回答了这些问题，水分就会被传递给花粉，花粉管就向下生长直到花柱。可能有不止一个花粉粒已经发芽，有证据表明花粉粒在当前存在一种对卵子的竞争，针对特定性状的选择可能在这一阶段发生。

当花粉管从花柱的末端伸入子房时，它会朝卵子的方向生长，并且释放两个而非一个精子。其中一个精子做了体面的事情，找到一个卵子并与其结合成真正的合子。这个合子长成胚胎，胚胎的发育随后受到抑制，正如裸子植物中发生的那样。然而，第二个精子不仅仅是一个备用精子。它需要找到具有两个单倍体核的雌配子体细胞。第二个精子与这个奇怪的细胞融合，所产生的细胞就是三倍体，即每个染色体都有三个副本。这个双核细胞可能被看作“世界上最重要的细胞”的原因在于，它继续生长为胚乳，而胚乳不仅是胚胎的食物供给，也是智人的食物供给。当我们食用由水稻、小麦或甜玉米制成的所有食物时，我们吃的就是三倍体的胚乳。

那么，开花植物的花粉究竟是如何进行精准移动的呢？好吧，在某些情况下，花粉并不能移动，它们通常在极为偶然的机会中利用风来移动。花粉很少利用水，但是如果能为动物提供适当奖励的话，它们通常会雇用动物。奖励可以是多种多样的。55 饲料（花蜜、淀粉以及花粉本身）、一张过夜的床、一张和别人过夜的床、一个产卵和养育后代的地方，这些都是植物所能提供的奖励。奖励的大小取决于来访者的体型，所以鸟或蝙蝠要比摇蚊需要更大的奖励。吸引传粉者非常重要，但动物都很直接，所以吸引动物也很简单。气味、形状和颜色都会被利用起来。有些植物只与一种传粉者陷入非常密切的关系，只要传粉者没有发生状况，这种关系就很明确。有些植物物种则不在乎是谁到访，只要有人来帮它们传粉就可以了。这将增加花粉落在不合适的柱头上的风险，但这株植物至少已经有人访问过。

我们通常可以将传粉者与花相互配对，而传粉综合征的研究也是查尔斯·达尔文实验工作的一部分，这给予了他最大的乐趣。他写下了大量关于植物采用各种计谋的内容。然而最近的研究表明，和植物生物学的许多内容一样，这种关系并不像人们想象的那样精确。有些没有特殊适应能力、看似通才的花仍然只接受一种传粉者的到访，相反，有些看似适应良好的花则被许多不同的生物访问。同样地，尽管我们有可能描述出完美的蜜蜂授粉花或者是鸟类授粉花，但当你寻找它们的时候，自然界中却很少有这样的植物。

因此，授粉，或者说未成熟雄配子体的分布，是种子植物生命中至关重要的阶段。它不仅有助于基因库的活跃，也有助于基因的分布。在非种子植物中，有鞭毛的精子必须承担起这一责任。由于这种机制已经运转了超过4.5亿年，因此我们很难将其视为一个问题。然而，还有另一个问题既不适用于有鞭毛的精子，也不适用于传粉者，那就是新胚胎离开亲本植物的传播。种子植物有种子，而没有种子的植物则有另一种方法，我们将在下一章中讨论这些内容。

第四章

# 植物的传播

如果你生活在水中，那么有两种方法活动：漂流或是游动。有些灰胞藻是游动者，有些红藻是漂流者（如多管藻），有些绿藻是游动者（如衣藻、团藻和石莼），而有些绿藻是漂流者（如水绵和轮藻）。这里所说的游动是指孢子（例如石莼）或整个生物体（例如衣藻和团藻）的游动。精子的游动不算数，因为这不是营养扩散，而仅仅是雄性生殖系的遗传扩散。精子不能发育成独立生存的生命体。另一方面，石莼的游动孢子则能够长成一株独立生存的植物。与陆生植物相比，生活在水中的植物的传播是否重要呢？传播对陆生植物而言是否重要呢？这些问题的答案在某些社会中似乎根据人类的状况而充满了主观色彩。据推测，在某些地区，人们认为儿童会离开家乡，而在其他地区，大家期望他们加入家庭团队并支持整个族群。植物没有情感，传播机制只有在能够增加植物基因进入下一代的机会时，才会被植物选中。

让我们回到以上问题上来。植物的传播是否重要呢？水环境要比陆地更加均一。陆地环境中存在的意外障碍极少出现在水中；水中存在梯度变化，但这些变化往往是渐进的。在这种环境中，游动是很有用的，因为这能使年幼的植物远离同物种的其他成员，因为这些成员对其造成的竞争要比任何其他的有机体更为激烈。然而，同样适用于陆生植物的是，如果亲本植物在某个地区生长良好，那么子代为什么要冒险离开这里呢？或许，这就是多达80%的陆生植物物种选择仅仅把种子撒在根部，而并不试图促进传播的原因吧。与水生环境相比，陆地在空间上具有异质性，因此我们或许以为有更多的陆生植物会在传播机制上下功夫。然而，除了空间异质性外，陆地在时间上也可能非常异质，因此我们需要同时考虑植物在时间和空间上的传播，因为植物可以以动物只能梦寐以求的方式进入假死状态。



让我们先来看看苔类植物和藓类植物。这些植物产生单倍体孢子，孢子再生长为独立生存的配子体。我们不仅会在地面上看到这些植物，也能在树上、高高的屋顶和墙上，偶尔也能在旧路虎的车窗管道里看到它们。孢子足够轻，可以被吹到任何高度，而且多亏了它们的孢粉素外壳，孢子也能在旅途中存活下来。蕨类植物在一定程度上也是如此。有许多蕨类植物并不生活在土壤中。有些附生蕨类植物生长在树枝外，有些蕨类植物是自由飘浮的，有些则生长在墙壁上的石头缝里。对于附生的蕨类植物物种而言，孢子必须能够到达它们赖以生长的树枝上。这同样适用于附生开花植物，比如兰花、多种凤梨科植物以及天南星科的部分成员，但我们稍后再介绍这些开花植物。所以从最随意的观察来看，苔类植物、藓类植物和蕨类植物在生存和传播上似乎没有困难。此外，孢子非常小，因此不会成为颇具吸引力的食物来源。种子却并非如此，那么种子有什么特别之处

呢？为什么进化会偏爱于生产种子这种消耗能量的结构，而种子又成为许多生物最喜欢的食物呢？ 58

种子生物学的规则即便存在也非常少。在理解种子生态学之前，我们仍要进行大量的研究。有一个问题是，种子植物至少有35万个完全不同的物种。同一个属内的物种在种子生物学上可能有所不同，甚至同一物种的植物在不同的地区可能也有所不同。例如，土耳其松（*Pinus brutia*）的生长范围从希腊克里特岛直到希腊北部。在其生长范围的南部，它从未经历过霜冻，而在北部冬季的夜晚，气温常常降到零度以下。生长在南方的土耳其松的种子不会休眠，它们在秋雨来临时就会萌发，就像世界上地中海气候地区的许多植物一样。另一方面，北方的土耳其松种子不想在严冬来临之前发芽，因此它们的种子需要在来年春天生长之前进行冷藏。它们仍然是相同的物种，因为气候驯化与物种形成并不相同。

有一件事我们可以不用担心矛盾地说明一下，那就是种子要比亲本植物小得多，而且它们能够耐受足以杀死亲本植物的严酷条件。正是这种能力，也许要比其他任何能力更能影响植物产生种子的各种策略。对于那些一生中只开花一次、因而也只产生一批种子的植物物种（一次结实植物）而言，至关重要的就是至少要有一颗种子存活到开花的时候，并将基因传递给下一代。本章可能会多次提到这句话，但它经得起重复：绝大多数种子都无法成长为能够产生更多种子的植物。如果是一棵橡树，能够在多个季节产生种子，那么这不是个问题，但一年生植物仍然只有一次机会。

35万个种子植物物种中的每一个都与其他物种有所不同。

同一物种中的每个个体在其一生中都要做出选择，从而导致了更多的多样性。生物学家们有时试图把生物学简化为资源配置问题，以便对生物体的习性和行为做出合理的解释。因此，当植物产生种子时，它们必须平衡植物体的生长需求与种子生产的能量需求。我们可以很清楚地发现，产生种子需要植物付出很大的代价，因为在那些雌雄异株的物种中，与雄性植株相比，雌性植株往往体型更小、寿命更短。

你会选择产生许多小种子，还是产生较少的大种子？如果产生大量的种子，这会增加你的生存概率吗？事实上，一次结实植物往往会产生非常多的种子，这意味着植物生产的种子越多，其生存概率越大。你会选择用脂肪还是碳水化合物作为种子的能量供应呢？脂肪的制造成本更高，但每克脂肪所含有的能量更加丰富。植物将2.3%～64.5%的资源投入到种子生产中。种子生物学很少有规则可言。

另一个问题就是，植物需要多长时间才能成熟。与动物不同的是，同一物种中的性活跃植物在体型大小上可以也确实有很大不同。如果遭到强烈的破坏，植物往往会在下一次灾难性变化来临之前迅速地完成生命史。在环境比较稳定的地方，种子生产可能是稳定不变的。例如，与无花果小蜂有着独特关系的无花果树全年都会产生花朵来满足小蜂的需求。另一方面，山毛榉树则表现出一种不同的大量结实策略，也就是大约每7～10年才产生一大批种子。人们认为这种策略的优势在于植物产生的种子数量要远大于自然捕食者能够消耗的种子数量。另一方面，橡树每年都产生许多橡子，其中许多橡子被松鼠藏起来后又被它们遗忘。这对橡树来说似乎是个好消息。

图9　巨子棕（*Lodoicea maldivica*）具有世界上最大的种子。它只生长在塞舌尔的普拉兰岛上

种子的大小差别很大。兰花的种子极小，几乎像是灰尘；而巨子棕（又称海椰子）的种子足有两个澳式足球那么大；也就是说，种子的重量有七个数量级的差异，但兰花种子和巨子棕种子行使着同样的功能，二者都有潜力长成成熟的开花植物。即使在同一株植物中，种子的大小也能达到三个数量级的差异，而同一个果实内的种子也能有所不同。有一条经验法则似乎是，生活在阴凉处的植物的种子往往具有更大的尺寸上限，这可能是因为更大的种子能令它们在能量水平较低的环境中生长。种子的大小似乎与土壤的含水量和养分状况均无关系。这可能只是因为，植物通过生产大小不同的种子来规避风险、增加生存机会。同一尺寸的种子并不适用于所有情况。



在前一章中，我们看到有许多植物依赖陌生物种来传播它们的精子，而那些不信任动物的植物真的就是抛掉谨慎随风飘荡。我们知道，生物或非生物授粉方式都无法完全令人满意，因为植物的人工授粉几乎总是要比大自然的方法更加成功。在这方面，开花植物的一些科尤其不可救药。在山龙眼科中，尽管受到传粉者访问的花有很多，但最多仅有7.2%的花能成功授粉。这可能是因为它们选择的传粉者如今已经灭绝了。（植物物种在与动物物种进行联合经营时普遍存在一个问题，那就是植物物种的进化周期大约是动物物种的30倍。因此，如果你打算用动物来传播花粉或种子，你就必须做好反复更换供应方的准备。）

我们在植物繁殖中见到了阿利效应[①] 的一个很好的例子，也就是说一个物种的成员最好是待在一起——这是科学的安全数



① 阿利效应是物种生态学的重要原理之一：当种群密度低于某一阈值时，物种将会灭绝。

量原则——因为植物和花的数量越少，成功的授粉就越少，因此产生的种子就越少。在你开始认为其中即将有规则诞生之前，必须记住，虽然更大的花序群会吸引更多的传粉者，从而产生更多的种子，但它们也会吸引更多的食花动物、更多的种子采食者和更多的果实采食者。因此种子没有最佳尺寸；最佳尺寸是依据时间和空间而特定存在的，所以最佳策略就是变异。

查尔斯·达尔文不仅对授粉感兴趣，而且对物种的地理分布也十分着迷。他曾在其故居里进行了一些精妙的实验，以探索陆生植物的种子能否在海水的盐度下存活足够长的时间，使其能够从一块大陆漂移到另一块大陆。他认为鸟类的足是另一种远距离传播的方式，特别是传播到海洋岛屿，这是孩子们在学校自然课上最早学到的内容之一。老师会向孩子们展示拉拉藤属（*Gallium*）和起绒草（*Dipsacus fullonum*）的果实是如何附着在动物的皮毛上的，而且由于这是我们最早了解到的内容，所以我们往往认为大多数植物都是这样传播种子的。事实上，只有不到5%的物种会依附在动物体表旅行。这些物种生长低矮（即动物的高度），并且来自多种不同的栖息地。在我们介绍过美国悬铃木的“直升飞机”种子后，风的传播也被认为是很常见的，但是很少有结构能改善种子的横向运动，它们只是能令种子下落得更慢，从而使水平向的风产生更大的影响。如果风力很强，便能造就非凡的传播，但是人们已经证明，被动物藏起后却遗忘的一批种子要比被风吹散的一粒种子距离母体植物更远。因此，动物皮毛和风的确有助于种子的传播，但只有少数物种会利用这些传播方式。一些亲本植物通过弹道机制将种子物理喷射出去。其中最壮观的可能要数喷瓜（*Ecballium elaterium*）了，

63 当果实从茎上脱落时，它会将种子以废气流的形式释放出来。这种传播方式极其有效，每粒种子都能在迷你果实的短暂飞行中被排出。

除这两种方式之外的主要传播方式就是种子被动物吃掉，然后被动物排泄到距离母株有一定距离的地方。这是一种冒险的策略，因为动物（通常是哺乳动物或鸟类）出于营养价值而吃掉种子，所以它们会碾碎并杀死种子以便提取食物。有趣的是，那些善于在山羊肠道中存活下来的种子同时也能在土壤种子库中保存多年。有一种可以降低被动物消化的风险的方法就是在种子上覆盖一层泻药；这在由鸟类传播的物种中更为常见。对植物而言，这可能是一种很好的策略，但这确实给鸟类传播的研究造成了困难，因为很难为这种随机的传播绘制地图！就像传粉者和花可以配对一样，种子和动物也可以配对。鸟类容易被无香味、色彩鲜艳（通常会结合红色、蓝色和黑色）的种子所吸引，而哺乳动物则喜欢有臭味、味道鲜美但颜色暗淡的种子。动物可传播有活力的种子的实际证据十分薄弱。在两项研究中，研究人员分别对40 000份鹿的粪便和1 000公斤犀牛粪便进行了检查，并没有发现有活力的种子。另一项研究中，在40 025粒被朱雀食用的种子中，只有7粒幸存下来。



把动物肠道作为传播媒介不切实际而且愚蠢，很多种子和果实有毒可证明这一点。正如植物生物学的许多方面一样，种子或果实因具有毒性而受益有许多原因，而这些原因绝不是互相排斥的。一个美味但含有泻药的果实将确保种子能够迅速且安全地通过消化道。催吐剂可以确保种子在进入消化道之前被再次迅速且安全地排出。如果种子或果实有毒，那么它可能会

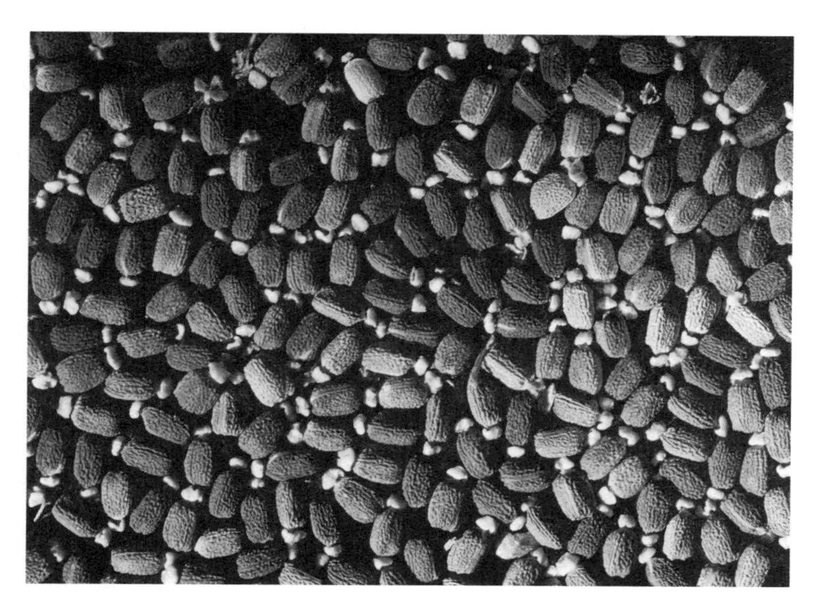

图10 黑叶大戟（*Euphorbia stygiana*）的种子，显露出油质体

杀死动物，借助腐烂的宿主给幼苗提供营养。毒素可能会保留其中，抑制种子的萌发，也就是生理休眠，我们在后文中会更详细地介绍。毒素可能对种子采食者有毒，但是对果实传播者无效。其中一个很好的例子就是英国紫杉，这种植物红色的假种皮可以安全食用，但种子是有毒的。最后，毒素也可能起到对病原体的防御作用。 64

蚂蚁是在种子传播中至关重要的一类动物，特别是在地中海型地区。经由蚂蚁传播的种子含有油质体。这种脂肪组织通常比种子小得多，是蚂蚁的一种致醉物质，只要它刺激蚂蚁拾起种子并把种子带回巢穴，在进入蚁巢之前，蚂蚁会咬下油质体，把完整的种子留在外面。蚁巢周围的土壤往往比附近的土壤养分含量更高，而且种子通常被带到地下几英寸的地方，在那里种子不仅能免遭其他动物的侵害，而且还不会受到这些地区常见火灾的有害影响。许多植物物种都依赖于蚂蚁的这种保护，因此当非本地的蚂蚁物种驱逐出本地物种时，这些植物就会很容易遭受地方性灭绝。



关于非本地物种的问题，我们将在第七章中更详细地讨论，但现在我们所讨论的问题与之相关，因为植物生长在这里而不生长在那里的主要原因是它们无法到达那里。然而，人类已经无意或有意地打破了每一道传播屏障。农民通过受污染的种子袋、肥料和牲畜来传播种子——已有证据表明400只绵羊每年能移动800万粒种子。园丁已经把成千上万的物种远远移到了它们的自然生长范围之外。最后，林业工作者把更具生产力的树木引入许多国家。我们不能低估这些非本地物种可能造成的损害。



种子传播的“益处”有多种原因。第一，亲本植物将成为害虫和疾病的向往胜地，极少有植物疾病能在种子内部或表面存活。这种掠食者和疾病的解除是植物在新引入的国家大肆扩增的原因之一。第二，传播种子可以减少随机发生的或者突发且不可预测的灾难所造成的风险。第三，它减少了植物的亲子竞争（此处将植物学更加拟人化）。第四，它增加了植物找到“安全地点”的机会。然而在现实情况中，超过80%的植物不会做出任何努力去散布它们的种子，所以除非植物需要进行快速迁移，否则没有对种子传播的适应性并无不利之处。如果气候变化得很快，那么在接下来的几十年和几个世纪里，植物可能会非常积极地选择快速迁移。

种子是植物的传播单位，而由于种子在土壤种子库中的生存能力，我们可以说种子在时间上被传播了。我们很难对在土壤种子库中会找到哪种植物和哪类种子做出概括，其中有部分原因是很难找到种子。不过，一般来说，小种子在土壤中存留的时间更长，因为它们对掠食者的吸引力较小，而且比大种子更容易落入土壤的缝隙中。如果种子足够小，土壤可以容纳许多种子，目前已有的最高纪录是每平方米土壤中含有488 708粒种子。种子在土壤中的最长存活时间纪录是由神圣的莲花（*Nelumbo nucifera*）种子所创造的，这颗种子在埋藏了1 288年（前后可能相差250年）后仍能萌发。在英国，W.J.比尔曾于1879年发起了一项实验。在土壤中埋藏了120多年后，人们仍然可以从中挖出有活力的种子。

66

图11　莲的种子可存活超过1 000年

土壤种子库中的种子可能没有休眠，它们可能只是身处的萌发条件不对。在这种状态下，我们认为种子处于长期的静止状态，而对于埋在地下的种子来说，光是它们所缺失的条件。为了存活下去，种子可能是干燥的，但有些种子在吸水的情况下能存活得更好，因为在这种条件下它们能够更好地修复膜和DNA上的日常损伤。然而，种子埋得越久，就越有可能死于病原体、掠食者或者放牧后的过早萌发。持久生存的能力似乎属于某个物种的特性，而非属于某个物种中的个体。持久性还与完全无67法将植物培育出下一代的潜在风险有关。这意味着，像林地这样的稳定群落中的植物在土壤种子库上的持久性不如地中海型生境中的种子。

我们已经提到过休眠的复杂概念，还需要正确地理解它。种子不会在一大包种子中发芽，这是因为环境不满足萌发的基本要求。这些种子处于静止状态。如果某时某刻的环境是错误的，那么它可以阻止种子发芽。休眠是种子的一种状态，而非种子所处的环境。如果种子在播种后的一个月内没有发芽，那种子要么是死的，要么处于休眠状态。如果种子处于休眠状态，那么休眠必须被打破，种子才能发芽。休眠的作用是控制发芽的时机，使得幼苗获得最大的生存机会。在环境条件一致且可预测的生境中，休眠就不那么重要，因为这些条件本身就会阻止种子萌发。热带潮湿林地的大多数物种不会产生休眠种子。

休眠有三种基本类型。形态休眠是指种子中的胚胎在脱落时尚未成熟，例如兰花的种子。物理休眠是指坚硬的种皮可以阻止水分的吸收并保持胚胎的干燥，例如豆科植物的种子。生理休眠是指种子需要一些化学变化，例如蔷薇科植物的种子需要冷藏一段时间。前两种休眠类型是不可逆的，而生理休眠则是可逆的，因而也具有一定的灵活性。形态休眠和物理休眠从不同时发生；形态休眠和生理休眠通常一起形成形态生理休眠（MPD）；而物理休眠和生理休眠极少同时发生。

休眠会把植物困在特定的生境里，一旦气候变化，植物就极其容易受到影响。然而，亲本植物所经历的条件会影响休眠，因此同一个物种的植物可以表现出不同的休眠状态，这取决于它们的生长环境。例如，生长在克里特岛的土耳其松无须休眠，并在雨水来临的秋天发芽，而生长在希腊北部的土耳其松的种子则需要一段时间的冷藏来阻止萌发，直到冬天结束、春天到来。68

一种普遍的观点认为，坚硬种子所引发的物理休眠是由一段时间的磨损或者微生物的攻击所打破的。目前没有证据可以支持这种观点。这种休眠的打破要受到更多的调控，由各种不同的结构提供温度控制的单向阀，以允许水分的进入。这种策略已经在几个不同的植物类群中发生独立的进化。另一种已经发生多次进化的策略是寄生，寄生物种的种子有一种非常巧妙的感知功能，当寄主植物就在附近时便可以察觉到它们的存在。



吸水是种子萌发的第一阶段。随后便是呼吸作用的迅速增加和食物储备的调动。胚胎开始生长，当幼根长出来时，种子已经发芽，一切已不能回头。许多因素都能刺激萌发。昼夜间的温度波动是种子感知上方树冠空隙的一种方法。同样，感知落在种子上光的质量和数量的能力将使种子萌发的时间与一年中特定的时间、特定的埋深或阴影程度相协调。光敏感是由种皮引起的，值得记住的是种皮是由母体植物所提供的。这是亲本所经历的环境条件能够同时影响种子的休眠和萌发需求的一种方式。

可利用水分是种子萌发的关键条件。并非所有的种子都是干燥并且能耐干燥的。至少有7.4%的物种所产生的种子并不遵从这一规则，这意味着它们无法忍受土壤的干硬。土壤营养水平对某些种子的萌发有着重要的影响。杂草需要高水平的氮。树冠间隙的形成通常伴随着氮的增多，因此光照、温度和土壤养分状态可以相互联系起来。打破休眠并促进发芽的一种普遍机制是烟熏，人们甚至在目前并不生活在火灾常发生境的物种中也发现了这种现象。

通常认为，优势植物的落叶层能抑制种子的萌发。尽管这种抑制后代的想法很有吸引力，但这种化感作用从未在自然系统中得到证实。落叶层可以为种子萌发提供一种机械障碍，或许可以将需要光的种子维持在黑暗状态中。落叶层对种子萌发和幼苗形成的影响是一个经典的生物学难题，因为落叶层在为种子萌发提供良好苗床的同时，也可以成为蛞蝓和蜗牛等小型食草动物的栖息地。

因此，种子萌发受到温度（实际温度和相对温度）、水分有效性、营养水平（尤其是氮含量）、火灾以及土壤种子库的存在等因素的影响。人们猜测，所有因素都会随着气候的变化而变化。气候变化对种子行为的影响可能意义深远。

种子即便已经萌发，也仍然处于危险之中。研究表明，超过95%的幼苗可能被吃掉或以其他方式被杀死。这个数据还排除了所有被吃掉或是落在不适当位置的种子。这些损失可能都不是特别引人注目。种子无法成长为成熟植株的最大单一原因是缺少合适的生长间隙。约翰·哈珀引入了“安全地点”的概念，即满足种子萌发和幼苗成功形成的所有条件的地方。这些地点可能非常小，或许只有1%的种子能够找到一处安全地点。土壤70必须处于合适状态，菌根真菌必须存在并将在7～10年内与之形成关联，并且光照水平必须适宜。

种子生物学可能看起来并不精确。我们如今对种子的了解要比20年前多得多，但仍然没有总结出任何规则。这是不是因为种子生产是一种进化上有风险的策略？胚胎在任何生命周期中都属于一个脆弱的阶段，但是对所有的亲本来说，在释放胚胎的同时吸引捕食者为其提供食物似乎都是一种奇怪的行为，但它却排除万难做到了这一点。芬纳和汤普森曾经明智地提出，世界上存在“因未加选择而导致物种多样性的再生抽彩”。这种抽彩造就了40多万种植物。我们该如何梳理这种多样性来帮助我们理解生物学呢？第五章告诉你答案。71

第五章

# 植物的多样性

给植物命名是最古老的职业。如果你翻开《创世记》的第二章第19节，上帝把他创造的作品带给亚当，表示自己已经做完分内之事，现在轮到亚当来为它们命名。不管你如何看待这段历史叙述的真实性，它着实表明圣经的作者已经意识到，如果他想要继续讲述故事，植物就必须有名字。亚当的任务完成得很快，因为仅仅三节经文过后，上帝就把他当作了闲暇时的玩物。亚当能够很快地完成这项任务是因为一切都是从零开始，而且世界上只有他一个人。

如今的情况要复杂得多，因为许多土著居民都会给植物命名，并且互不参考。所以白金汉郡的牛眼菊[①]被奇尔特恩丘陵另一侧的牛津郡称为月亮菊。在德国，白睡莲有108种名字。在英语中，“百合”（lily）这个词本身被广泛应用于诸如睡莲、马蹄莲、铃兰和白百合花等多种植物。其中只有白百合花才是真正

① 指菊科植物滨菊（*Leucanthemum vulgare*），又称法国菊、西洋菊等。——译注

图12　依据不同的生长地区，牛眼菊有许多不同的英文名字



的百合。这些俗名有一个优点，它们都是当地语言，因此更容易被人记住。这一点非常重要，比如颠茄在英语中就被叫作“死亡夜影”。然而，这个名称在法国毫无意义，在那里同样的植物

72 有着不同的名字。我们所面临的问题是，在人们到处旅行、相互交谈的世界里，植物必须有一套世界公认的名字。伟大的全球植物名称编目员林奈曾说过："没有永久的名称，就没有永久的73 知识。"

多亏了林奈（以及其他分类学家），我们真的拥有了永久的名称，这就是以拉丁语（或希腊语）命名的双名法学名。拉丁语和古希腊语的伟大之处在于它们是与政治无关的语言，因此不会有人出于民族主义的理由进行控诉。它们也不会像其他语言那样继续进化。如果你懂一些拉丁语或希腊语，那么你就可以开始欣赏植物命名的描述性质了。朱缨花（*Calliandra haematocephala*）是一种豆科植物，它的花朵紧凑生长成半球

图13 朱缨花的花看起来是由许多雄蕊构成的，因此它的拉丁名意为"美丽的雄性部分"

状。每朵花表面上看起来都是由雄蕊构成的，所以整个花序看上去就像一个粉扑，这也是它常用名的由来[①]。朱缨花的拉丁语学名指“美丽的雄性部分”，这和粉扑花一样好记，但这也确实使卡琳德拉（Calliandra）成为一个不适合女孩的名字。人们也确实有些同情中国人，因为许多中国植物都是以外国人的名字命名的。例如，十大功劳（*Mahonia fortunei*）是以宾夕法尼亚州的一名园丁和一位苏格兰植物标本采集者的名字而命名的。 74

现在，学术界已有条例规定了植物的合法命名。首先，你要检查这种植物是否已有名称。在你确定它是一种新的植物物种后，你可以用拉丁语为其命名并对其进行描述，虽然2011年的国际植物大会决定拉丁语描述不再是强制要求。其次，你需要制作一份干燥的压制标本，并将其放置在可获得永久保存的标本室里。最后，当你在同行评议的期刊、博士论文或类似论文中发表植物的名称及其描述时，你需要说明该植物标本的存放地点，这样植物学家就能在未来的至少350年内来查看你所命名的实际植物。尽管植物命名具有相关法规，但人们对已发表名称的审查不够充分，因此，每个开花植物物种平均有两三个合法名称。这些名称被称为异名，而使用最为广泛的名称通常会成为公认的名称。

如果和来自世界各地的植物学家交谈，你会发现许多命名系统的存在，而这些命名绝对不是像儿童洗礼名一样的随机分配。例如，在中国，人们把所有的木兰称为辛夷花。在新西兰，朱蕉（*Cordyline fruticosa*）在毛利语中被称为Ti植物，其不同的



① 朱缨花属植物又名粉扑花、红绒球等。——译注

品种则被称为这个Ti、那个Ti和其他Ti。双名法是许多方言命名系统的共同特点。当人们为自己在周围看到的植物命名时，会凭直觉把相似的物种归到同一类。我们能够在被认为是现代植物学开端的一本书中发现这一点，那就是泰奥弗拉斯托斯在大约2 300年前出版的《植物探究》。

泰奥弗拉斯托斯记录了他对植物的观察，并在这个过程中根据植物的外形对其进行分类。例如，他这样描述一束植物：花朵都从同一个点长出，有许多全裂叶在花柄的基部紧紧地围绕茎而生长。这些都是伞形植物或者说是伞形科（Apiaceae）的成员。这本书在1644年仍作为教科书而出版，并且还有翻译本和复印本可供使用。植物学的研究一直持续到希腊和罗马

75

图14　豕草的头状花序。豕草属于一类叫作伞形科的植物，而伞形科植物自公元前3世纪就被人们所认识

帝国，在公元1世纪中叶，狄奥斯科里迪斯出版了《药物论》（*De Materia Medica*）。这部作品直到17世纪都还在使用中。它介绍了狄奥斯科里迪斯在药物中所使用的植物，但这些植物是按其形态特征和药用特性而进行分类的。（在后文中我们将看到这两种特性经常融合在一起。）

随着欧洲文艺复兴的发生和旅游业的指数增长，人们对植物本身重新燃起了兴趣，而不仅仅是为了它们的实用价值。许多植物志是在15世纪及以后编纂出来的，这些书是功利主义分类的极佳案例。我们现在仍然能够看到这些非正式的类别。在花园中，我们根据植物所需的条件（岩石花园、水上花园、林地花园等）或它们的颜色（希德科特庄园的红色边界或者西辛赫斯特城堡的白色花园）对其进行分类。我们有蔬菜园、药草园和果园。在其他社会中，我们可以看到食用植物、纤维植物、有毒植物等类别。这些分类显然是有帮助的，而且永远不会消失，但它们并非包括所有植物，并且也不是唯一的分类方式，因此一种植物可能会出现好几次。

76

早期的植物学家们，例如瑞士的博学家康拉德·格斯纳（1516—1565），开始描述他们在周围看到的所有自然世界。格斯纳是一位一丝不苟的工匠，如今保留下来的注释画稿和木刻展示出他对于细节异乎寻常的关注。与当时的其他植物学家不同的是，格斯纳采用了他能观察到的每一个特征，而非基于单个特征来对植物进行分类。这种对于发现某种本质特征的痴迷与亚里士多德哲学有着直接的联系，并且一直在植物分类中保持着卓越地位，直到约翰·洛克（1632—1704）引入了“观察是通向知识之路，我们生来一无所知”的思想。约翰·雷（1627—

1705）就是深受洛克影响的人之一。

分类学（taxonomy）就是把生物进行分类的科学（*taxis* 在希腊语中意为顺序或排列），而约翰·雷正是现代分类学之父，也是英国诞生的第二位伟大的博物学家。他同样喜欢摆弄植物、动物和岩石。他对那些已经不复生存的动物化石深感困扰——他称之为“大自然的游戏”。然而，尽管雷有着强烈的宗教信仰，他还是制定了分类学的基本规则，并自此以来构成了所有分类学的基础。

雷是第一个定义“物种”概念的人。这在如今看来似乎毫不惊奇，也不值得注意，因为我们在日常交流中就会使用“物种”这个词，因此假定这个词语一直具有它的定义和含义，但在雷的时代并非如此；事实上，即便在今天，“物种”这个词仍然有20多种定义。物种是分类学的基石，也是生物学的通用概念，因此它的定义至关重要。雷的理念认为：

图15　斯塔福德郡的米德尔顿大厅，约翰·雷离开剑桥后曾在此工作

> 在我看来，确定物种的最可靠的标准莫过于使其在种子繁殖的过程中保持自身的区别性特征。因此，无论个体或物种发生什么变异，如果它们起源于同一种植物的种子，那么它们都是偶然的变异，而不应被区分为一个单独的物种……；一个物种绝不会起源于另一个物种的种子，反之亦然。

因此，物种是一组相似的个体，它们可以自由杂交并产生与亲本相似的后代。这一生物学定义至今仍被许多人接受，并在许多考试中得到引用。当你试图将这种定义应用到兰花、蒲公英或是其他植物上时，其中就存在着明显的问题，因此一个更为简单可行的定义就是："物种是一组个体，它们共有一套独特但可以复制的特征。"

因此，雷认识到，如果想要维持某个物种内部的自然变异，这个物种就应该通过种子进行繁殖。这是真正的预言，因为这距离孟德尔发表他的遗传学（以及变异）研究还有200年，距离英国皇家植物园邱园确立千年种子库计划并以种子的形式来保存世界上25%的植物物种还有300年之久。然而，雷对分类学的最大贡献来自他的主张（遵循洛克的原则和格斯纳的实践），即如果想达成自然分类，你必须采用你观察和/或测量到的**每个**特征。你不应该无缘无故地忽视任何特征。他的观察受到当时的显微镜分辨率的限制，但他的研究代表了植物分类学的一个转折点。（值得注意的是，雷及其同时代科学家的**自然**分类亦反映了**造物主**的计划。）

他摒弃了最早把植物划分为草本和木本的分类方式。他声明，所有的开花植物都专属于两大类植物的其中一类，而他将这两类植物称为单子叶植物和双子叶植物。他将单子叶植物的特征描述为：种子具有一片子叶；花部的基数为三；叶片具有平行叶脉；茎内的维管束组织呈分散排列；根为不定根。另一方面，他将双子叶植物的特征描述为：种子具有两片子叶；花部的基数为二、四或五；叶片具有主脉和横向分支构成的网状叶脉；茎中的维管束组织沿着茎的外缘呈环状排列；根是永久性的，具有一个存留的主根以及许多侧根。直到1998年，单子叶植物和双子叶植物的区分在植物分类学中一直存在，我们必须提到的是，由约翰·雷所定义的单子叶植物类别仍然存在。雷采用他所能观察到的每一种特征的进一步结果就是，他也开始把植物归入我们至今仍在使用的科这一分类单位中。例如，紫草科就是雷所描述的一个类群，尽管他没有采用“科”这个术语。

图16　棕榈的茎展示出单子叶植物典型的维管束组织

约翰·雷绝不是当时唯一的分类学家。他定期与巴黎植物园的约瑟夫·皮顿·德·图尔纳弗（1656—1708）通信。作为一位分类学家，图尔纳弗是以草本植物和木本植物进行分类的，但是通过观察花的结构，他在物种的上一个分类单位层面（也就是属）进行了大量的整理。这为卡尔·林奈（1707—1778）引入针对每个物种的通用拉丁语双名创造了条件。在实地调查中，林奈发现这些冗长、详细、描述性的多词学名非常烦琐。他建议保留图尔纳弗提出的属，但要把其余的名称缩减为一个希望能有助于识别或记住这种植物的昵称。1753年，他出版了《植物种志》（*Species Plantarum*），这标志着合法学名的开始。他用一个拉丁语双名以及自泰奥弗拉斯托斯和狄奥斯科里迪斯以来就应用于该植物的所有拉丁语多词学名列出了他所知晓的每一种植物。





有趣的是，在林奈的四本自传中，他认为1753年出版的《植物种志》不应算作他的遗产。对大多数科学家来说，这足以成为一个人的遗产，但林奈确信自己已经在1735年出版的《自然系统》（*Systema Naturae*）中所描述的性系统分类中找到了完美的植物自然分类。他完全按照植物所拥有的雄性和雌性器官数量来进行分类。他的著作非常生动有趣，也许他的出发点是让人同时受到启发和震惊。他的性系统分类从来没有被社会普遍接受，但如果你想鉴定一种植物的话，这种分类方法是有用的。然而，这的确表明将植物分类成木本植物和草本植物的方式已经不复存在。

林奈在二十多岁的时候就想到了采用这种方法对植物进行分类。他试图把植物从上到下进行分类。这时在你检查面前的所有物种之前，你要使用的特征就已经被定下来了。21世纪的大学生也经常在做同样的事情。另一方面，约翰·雷把他已知的所有植物都用非常完整的描述归类到物种，当你这样做的时候，植物的属和科几乎会自动选择植物；所以自下而上进行分类会更好。

我们当前的分类系统的下一步进展并不是由林奈，而是由安托万·罗兰·德·朱西厄推动的，他将林奈、图尔纳弗和其他人总结的属归类到科。这些归类在1789年的发表标志着合法科属命名的开始。

随着19世纪的开始，植物界已经被划分为物种，物种归类为属，属归类为科，科归类为目，目归类为纲，纲归类为门，最后门归类到界。这种分类等级的层次结构是针对不断增加的数据量的一种清晰而简单的梳理方法。当植物猎人们开始把植物带回邱园、爱丁堡等地时，这些新物种就可以被插入到系统中。每个物种都有自己独特的位置，在每个层级中都能归到某一类群。这使得植物的识别更加容易：要么通过询问一系列问题，比如这种植物是单子叶植物还是双子叶植物，来缩小选择范围；要么通过识别植物所属的科来进行鉴别。在植物界中，科是一项极其实用的分类层级，因为兰科、菊科、豆科、伞形科等多种科都很容易辨认。

然而，随着很多人开始怀疑物种的恒定性，一场革命也正在酝酿之中。有一种信念认为，世界上的物种是按照造物主的计划而固定不变的，它们也不能发生改变。这又回到了约翰·雷对化石及其他事物的担忧上。林奈发现许多植物展示出的特征是另外两种植物特征的混合体。他经常给这些植物起一个

"*hybridus*"（杂交）的特殊绰号［例如，车轴草属植物（*Trifolium hybridus*）仍然生长在林奈和学生一起采集植物的路线上］。如果所有的物种都是上帝在创世记的第一天创造出来的话，又怎么会出现杂交物种呢？林奈很巧妙地解决了这个问题，认为上帝创造了属，而自然创造了物种。

作为牧师的儿子，林奈或许知道何时该低调行事。然而，巴黎植物园的让-巴蒂斯特·拉马克（1744—1829）等人开始公然考虑物种会随着环境的变化而变化。拉马克受到有些人不公正的嘲笑，但达尔文和其他人给了他应得的荣誉。他是那些踢了进化的马蜂窝、为进化论软化科学世界的勇敢者之一，该理论在查尔斯·达尔文的第一版《物种起源》中得到了精辟的概括。 82

这本书应该成为所有生物专业一年级学生的必读书目，直到这些学生认真读完这本书后（在关闭i-pod的情况下），学校才应允许他们开始学习学位课程。这个看似严苛的想法的原因在于，达尔文和雷一样，也是一位博物学家，同样在家里钻研植物、动物和岩石。他的写作风格至今仍是科学阐释的典范。"激发、联系、揭示"是保护教育的准则，这也正是达尔文所执行的。当你读到《物种起源》的第十三章时，你已经完全被达尔文的思维方式所吸引。

第十三章与我们本章所介绍的内容有关，因为达尔文在这部分内容中讨论了物种的分类。他认为，如果他关于生物在进化中会发生改进的观点是正确的，那么就非常有助于解释我们将生物划分成这种嵌套等级的层级系统的事实。《物种起源》中唯一的插图就是分叉树——系统发育学的时代已经开始。达尔文认为，物种可以归入属，因为它们在过去的某个时间点曾拥有一个**共同的祖先**，而你对共同祖先的溯源越深，你所处的等级层次越高——**系谱近似**。分类系统并没有证明进化已经发生，但是达尔文提出的进化系统解释了我们为什么可以按照当前的规律进行分类。突然间，**自然分类**的含义就发生了改变。分类学家不再试图揭示造物主的计划；现在他们正在重演进化史。这本身就是对达尔文理论的一大反对意见，因为尽管人们相信造物主是有计划的，但进化却没有计划，这让人们感到害怕。

83

随着19世纪的继续前进，邱园等地的植物标本室中的植物数量不断增长，分类也随之发生调整并得到完善。1862—1883年间，乔治·边沁和约瑟夫·胡克爵士出版了自己的植物分类著作。在开花植物中，他们辨别出了单子叶植物和双子叶植物。在双子叶植物中，这些植物首先被分成三大类：花瓣彼此分离的植物、花瓣融合成管状的植物以及看起来没有花瓣的植物。在单子叶植物中，这种划分则基于各种各样的特征，例如花朵是五颜六色的还是棕色膜状的、种子是不是很小等。边沁和胡克的分类系统在鉴别植物的目的上十分实用，但他们没有试图揭示进化的规律，尽管胡克经常与达尔文通信。也许因为进化论实在太有争议了。



《物种起源》的出版为系统发育学的创立打开了大门。德国博学者恩斯特·海克尔（1834—1919）常常被认为创造了“系统发育”（phylogeny）这个术语。系统发育是一种进化分类，并且以分叉树的形式进行绘制，每个末端分支则代表一个类群。希望有一天我们能为所有开花植物建立起系统发育史，那么这个分叉树将有大约35万个末端分支，而每个末端分支都代表一个物种。海克尔的分叉树是基于他和其他人利用日益精密的显微镜所观察到的形态而绘制的，所以他们所使用的证据从本质上而言与雷及其同事所能得到的证据相似。

在20世纪，一种新的可能性出现了。随着1952年染色体作用和DNA结构的发现，以及随后遗传密码的破译，人们开始怀疑事情的本质是否如同亚里士多德所写的一般。我们是否有可能找到一段对每个物种成员都独一无二的DNA？我们是否能够利用从两个不同生物体中提取出的DNA片段中的碱基序列差异来确定它们的系谱近似关系呢？事实上，这是两件不同的事情。前者是鉴别，后者是分类。在这两个问题都没有得到回答之前，另一项技术革新打破了分类学的现状，这就是电子显微镜的发明，它打开了一个装满新特征的盲盒。 84

在电子显微镜下，最美丽的植物结构之一是花粉粒外部的图案。这就是我们在本书中多次提到的孢粉素。用电子显微镜拍摄的照片不仅揭示了特定物种所特有的图案，而且还揭示出花粉粒有许多萌发孔，当柱头面给出水分后，花粉管便通过这些萌发孔进行生长。萌发孔的数目从一个到几个不等，但通常是一个或三个。这是一项非常重要的特征，因为它帮助我们解决了一个越来越尴尬的问题。

随着电子显微镜所揭示的结构以及DNA序列（即分子数据）的增加，规则发生了一个非常重大的转变，那就是曾经和现在的单系性原则。单系类群指的是包括一个共同祖先的所有后代在内的一个类群。换句话说，一个单系类群中的所有分类单元都拥有一个共同的祖先，而这个类群必须包括该共同祖先的所有后代。单系性可以应用于每个分类层级。所以，所有的生物体会形成一个单系类群，因为生命发生了一次进化。植物（按照本书的定义）是一个单系类群，而陆生植物、种子植物、开花植物、单子叶植物、兰花等等也都是单系类群。没有哪种特征的新证据能够动摇约翰·雷在300多年前所描述的单子叶植物的单系地位。然而，双子叶植物的地位却并不乐观。

85

当达尔文开始思考特定系谱的相关性时，他遇到了一个问题。尽管种子植物（裸子植物和被子植物）似乎是一个单系类群，也就是说种子只经历了一次进化，但是它们的共同祖先是什么样的？被子植物的起源又是什么？在写给胡克的信中，他将此称为一个“令人憎恶的谜”，而这个谜至今仍未解开。试图揭示真相的一种方法就是找出最早的开花植物的形态。在分子数据和电子显微图出现之前，植物学家们一开始认为木兰属于一个早期类群，因为最古老的开花植物化石看起来与现代的木兰非常相似，而且木兰是借由甲虫传粉的，而甲虫的出现要早于蜜蜂，差不多就是在这些化石植物还活着的时代。除了木兰，植物学家们还在检验睡莲、辣椒、月桂以及其他几类植物。这些植物都是双子叶植物。随后，来自花粉的证据干扰了研究过程。结果发现，人们认为这些在某种程度上有些奇怪的植物群体的花粉全部都只有一个萌发孔，这与单子叶植物类似。而所有确凿的双子叶植物的花粉都具有三个萌发孔。



当分子数据与所有的宏观形态和微观形态综合在一起时，我们可以发现开花植物是由单子叶植物和真双子叶植物（eudicots）这两个非常庞大的单系类群以及其他一些小类群所组成的。这种看似异端的提议在1993年首次发表时，人们认为这种分类过于激进，以至于参与研究的科学家有半数都撤回了论文中的署名。到1998年的第二版论文发表时，没有人再怀疑单子叶植物和真双子叶植物的存在，但其他类群仍然需要整理。

目前，学界认为35.2万个开花植物物种应分类为（大约）41目、462科、1.35万属。

1998年的这篇论文是植物分类学的一个里程碑。首先，这篇文章拥有数量庞大的作者名单，因为这是一项跨国合作，在名义上由皇家植物园邱园的马克·蔡斯领导。然而，论文的引用署名并不是“蔡斯等”，而是被子植物系统发育研究组的缩写“APG”。其次，这篇论文被《自然》和《科学》都拒稿了，只能在密苏里植物园的《年报》上发表。150年来最大的植物分类学论文在英国媒体中获得的报道要比在顶流科学期刊中的报道更多，这清楚地说明了分类学在科学机构眼中的知识深度。再次，APG表示他们无法对部分类群进行归类。这是闻所未闻的。先前所有的分类都是完整的，每个分类单元都可分配到每个分类层级的特定位置。自第一版APG系统以来，现在已经有了第二版（2002）和第三版（2009）。细节越来越清晰。开花植物系统发育的最低分支是来自新喀里多尼亚的无油樟属（*Amborella*）。位于最低分支意味着，这个物种与其他所有开花植物物种拥有共同祖先的时间比其他任何开花植物类群都要长。它们通常被称为残遗类群。下一个分支则是睡莲及其亲缘植物；再下一个是木兰藤目（Austrobaileyales），这个类群里包括最著名的八角茴香，而有望治疗2009年猪流感的药物达菲就提取自八角茴香。自这些分支之后，分类细节就十分含糊。木兰、黑胡椒、月桂和鳄梨似乎组成了一个很大的类群，但我们并不知道这一类群与单子叶植物、双子叶植物或是古怪的金鱼藻属（*Ceratophyllum*）和金粟兰属（*Chloranthus*）之间的关系。只有时间能证明一切，但达尔文那令人憎恶的谜至今仍是个谜。

86

在对全世界的植物进行了这么多分类研究后，人们可能会问“那又怎样？”。这仅仅是一个为植物收集者创造就业机会的方案吗？这种公认具有主观性和哲学成分的学科真的是一门

图17　木兰花似乎至少生存了1.15亿年

科学吗？后者很容易回答。分类学试图为随机进化建立排序。物种处于不断的变化中。一个物种与该物种的变种之间并没有根本区别（达尔文，1859）。因此，分类学就像把果冻钉在墙上 87 一样杂乱无章，但仍值得进行，原因如下。

第一，系统发育分类使我们能够在环境保护工作中做出客观的决定，我们必须在保护工作中将不足的资源分配到最需要 88 的地方。例如，有两处相似的植被区域，每个区域包含8 000个物种，但你只能保留其中一个区域。如果一处区域包含20目、100科的植物，而另一处区域包含10目、50科的植物，那么前者可能比后者在进化树中包含更多的物种，因此也更有价值。

第二，如前所述，进化树中的某些类群要比其他类群富含更多的医用有效物质。茄科（Solanaceae）植物已为我们贡献了阿托品和莨菪碱等。当人们发现加兰他敏在阿尔茨海默病治疗中的作用时，我们只知道它来自雪花莲。我们还知道，雪花莲的生长速度不足以满足人类的巨大需求。人们很快就在石蒜科（Amaryllidaceae）植物中发现了一种外形类似水仙花的替代植物来源。同样，尽管用于治疗卵巢癌的紫杉醇最初是从短叶红豆杉（*Taxus brevifolia*）中提取而来的，它并不是一种可持续的供应物，而我们已经在欧洲红豆杉（*Taxus baccata*）中发现了原料的替代供应物并且可以进行商业化的提取。在过去的一年里，三尖杉酯碱已获准用于白血病的治疗，而三尖杉酯碱最早是在红豆杉的亲缘植物三尖杉属（*Cephalotaxus*）中发现的。

但是药物并不是我们从植物中获取的唯一收益，那么植物曾经为人类做出过哪些贡献呢？ 89

第六章

# 植物对人类的贡献

据说，保护植物有三类宽泛的原因，可总结为3S、3P或4E。具体来说就是明智/利己/精神，或者务实/实用/哲学，或者生态/经济/伦理/美学（几乎都是这样的）。毫无疑问，植物（以及通常而言的生物学）提高了我们的生活质量。它们美丽迷人，对它们的栽培也被许多人认为具有放松心情的效果。植物在艺术以及景观设计中的作用不容忽视，但这些都是植物的个体方面，属于精神/哲学/美学的范畴。

那么，植物的重要性还有其他两大类原因。一方面，大型生态服务通常可分为四大主题：供给（如食物或淡水）、调节（如固碳）、支持（如提供传粉者）以及最后的文化（如徒步旅行）。在某种程度上，这些服务并不局限于特定的物种，因此我们可以认为，只要树木在生长，究竟是哪个树种在生长和吸收二氧化碳并不重要。据计算，这些生态系统服务的价值约为每年330 000亿美元。这是一项奇怪而乏味的计算，因为如果你的钱包里真的有330 000亿美元，你要到哪里去购买这些生态系统服务呢？此外，全球每年的国民生产总值只有18 000亿美元，所以即便有人出售生态系统服务，我们也无力承担。如果某件事物不能购买，那么它究竟是珍贵无价还是毫无价值呢？这也许解释了过去几个世纪以来人类对待生物学的傲慢态度。

保护我们的生物遗产的其他原因就是物种特异性。这些正是从草本植物中提取或是以其他方式直接获得的商品和资源。其中就包括了欧洲红豆杉这样的植物，在治疗乳腺癌的泰索帝合成药物中，它为我们提供了Bacattin III。

思考植物对人类的贡献可能是最近几十年来人类才会提出的问题，因为我们当中有一部分人正在逐渐远离人类充分利用的植物。在某种简单的层面上，植物已经并将继续为我们做出一切贡献，因为它们（迄今为止）利用太阳能制造一系列令人惊叹的化学物质的能力十分独特，其中有些物质是全英国的化学家都无法合成的。



在另一种非常具有生物性的层面上，植物创造了我们目前正在与至少150万种其他物种共同存活的栖息地。然而，智人与其他物种的区别之一就在于人类能够增加所在栖息地的承载能力。有人提出，从长期来看，可持续农业只能养活15亿人，但从中期来看，到2050年预计最多需要养活9.5亿人。如果没有农业，地球只能支持大约3 000万狩猎采集者。这是一个发人深省（或者仅仅是荒唐）的数字，因为其他64.7亿人无处可去。为了养活当前的人口，我们正在耕种地球土地总面积的四分之一，并且我们正在以这样或那样的方式消耗所有光合作用40%的产物。人们极为关注的是，如果我们不得不扩大种植面积，那么有许多其他物种将被赶出目前的栖息地。

农业的起源并不明确。直到20世纪末，传统观点一直认为人类于大约8 000年前在底格里斯河和幼发拉底河之间的地区将草类驯化，该地区因而被称为新月沃地。人们认为这里曾经种植了多达八种作物：一粒小麦、二粒小麦、大麦、扁豆、豌豆、亚麻、苦野豌豆和鹰嘴豆，以及随后出现的黄豆。大约与此同时，水稻在中国得到驯化，土豆在南美得到培育。在那之后，世界各地出现了更多的独立驯化中心，包括在中美洲地区用墨西哥类蜀黍培育出玉米。到底是什么刺激了人类同步但不协调地采用这种农业生活方式呢？被称为“新仙女木事件”的气候异常事件在传统上被认为是农业发展的催化剂。这种简单的观点目前正在受到挑战。

图18　甜玉米现在是三种主要作物之一

现在人们认为，在上一次冰期结束时（大约14 500年前），中东或许还有其他地区的气温升高到了与当前接近的数值。大约13 000年前，气温继而迅速恢复到冰河时期的数值。这就是所谓的波令-阿勒罗德间冰期（Bølling-Allerød Interval）。接着便是大约12 900～11 600年前寒冷但干燥的新仙女木事件，在这一事件结束时，气温上升到远高于冰期的温度，冰川最终退缩。在新仙女木事件结束之后，气候罕见地逐年稳定，因此人类不再以15～50人为一组的团队四处移动，他们开始住在逐渐由石头建造的定居点中。这些史前人被称为纳图夫人，他们生活在如今的以色列、约旦、叙利亚和黎巴嫩地区。随着气温的升高，橄榄、开心果、小麦和大麦开始在这一地区生长。这些植物已预先适应了这些新环境。

现在大约有60处地点曾经有纳图夫人居住过，但是很少有直接证据表明他们曾经利用过哪些植物。其中一处定居点位于叙利亚幼发拉底河流域的阿布胡赖拉。在新仙女木事件发生前后，这里留下了两组遗址。有人认为，正是寒冷干燥的天气迫使在这片地区居住的人类不得不种植在野外愈发稀少的谷物。人们在这处定居点的遗址中发现了九颗饱满的黑麦种子。部分研究人员表示，这很难构成完整理论的坚实基础，尤其是因为黑麦还未在其他地区发现，并且几千年来都没有出现过。这可能是因为，新仙女木事件结束时，二氧化碳水平增加了50%，这通过使农业可进行光合作用而发挥了作用。

关于智人对植物的控制，人们目前有一些共识。我们知道狩猎采集者曾食用野生植物的果实、根、种子和坚果。这其中就包括草籽。2009年，有证据表明，在105 000年前，人类曾在莫桑比克使用石器碾碎谷物，尤其是高粱。这一证据因工具中存在造粉体而得到了支撑。造粉体是植物制造和储存淀粉的细胞器。淀粉沉积的模式和造粉体的大小通常具有物种特异性。此外，栽培品种通常具有更大的造粉体。野生辣椒的造粉体长0.006毫米，而栽培品种的造粉体长0.02毫米。造粉体非常实用，因为它们能抵抗腐烂，我们不仅在早期的厨房用具中发现了造粉体，而且在沉积物中也发现了它。它们现在已经被用于推断南瓜、木薯和辣椒在美国完成驯化的年代。

奥哈洛二号遗址位于加利利的西南海岸，这里在23 000年前曾有人居住。人们在这处遗址中发现了橡子、开心果、野生橄榄、大量的小麦和大麦以及其他等超过90 000份植物残迹，但没有哪种植物看起来像是栽培品种，也没有证据表明谷物曾经被碾碎。



就这个故事而言，重要的是要考虑栽培和驯化的含义。前者仅仅是有意种植或保护野生植物。生长是从种子或插枝等繁殖体开始的，而保护则是因为人类需要某种物质而对植物进行保留和培育。我们通常很难证明野生植物的栽培。另一方面，驯化是很明显的。驯化有许多定义，但这些定义都包括一种永久的物理和基因元素的改变，从而改良植物使其更适合人类的需要。

94

这些改良被称为驯化综合特征。有些特征显然对种植者更有利。这些特征包括：紧密的生长；令植物更具适应力；植物成熟的同时可进行一次收获；季节性开花和/或发芽的缺失可实现一年中不同时间的播种；更大的谷粒以及更薄的种皮以增加适口性，并使得磨铣等加工程序更加容易。风味和营养价值的增加是消费者希望看到的特征，而当农业开始发展时，种植者和消费者就是同一批人（在世界上许多地方至今仍是如此）。

对于一些作物来说，它们在驯化过程中需要克服非常特殊的障碍。就无花果而言，在我们所食用的“果实”能够正常发育之前，花朵需要借由一种特定的小蜂来进行传粉。无花果的花被封闭在隐头花序中（也就是我们所食用的无花果的肉质部分，严格地说这不是真正的果实）。我们在蔬菜水果店买到的无花果是一种变异品种，它可以在无须传粉的情况下发育出肉质的隐头花序。这些单性结实的品种是非常古老的。在约旦的新石器村落吉甲出土的无花果残迹展示了一颗这种古老的单性结实果实中的非凡细节，这些果实的年代可准确追溯到11 400年前。

令种植者感到沮丧的一种特征就是植物成熟后容易掉籽。种植者想让植物一直把种子和果实挂在上面，直到他准备好收割为止。这意味着驯化品种的种子不具有平滑的脱落伤口，而具有锯齿状的伤痕或裂缝。这对古植物学家来说是一种非常有用的特性，因为人们可以在非常古老的种子残迹中发现这种特性。土耳其奈瓦里·科里定居点的遗迹中就发现了这种种子。这处遗址中发现的种子是在10 500年前收获的，它们具有锯齿状的脱落层，这表明种子并非自然脱落，而是在收获过程中被撕掉的。这是有关植物驯化最早的直接证据。

然而，我们不应假定从野生谷物采集到选育品种种植的转变发生在一年之内。同样来自奈瓦里·科里的证据充分表明，野生植物和驯化植物是一起生长的，也许是因为种植者无法阻止部分野生种子在收获前掉落，从而使它们年复一年地保存在土壤种子库中。现在人们普遍认为，驯化之前要经过多年的耕种，因此，从狩猎采集社会到农业社会的转变过程可能涉及四个阶段。第一阶段是狩猎采集者以小团体的形式四处迁徙，他们





95

的移动受食物供应和天气的影响，而这两种条件可能很不稳定。这些流浪者可能居住在洞穴里，我们已经在这些洞穴里发现了他们日常生活的碎片。在高加索山脉的洞穴中，我们发现了1 000多份亚麻纤维。36 000年前，曾经有人在这处定居点活动。这些纤维似乎经过染色，尽管对亚麻进行染色是非常困难的。

众所周知，尼安德特人和后来的智人已经认识到，利用火可以获得极大的优势。小面积的选择性燃烧促进了柔软、可口的植被的再生。木质组织几乎不可能被消化，但食草动物消化纤维素不成问题。这意味着，再生燃烧区域可以吸引动物，而在未燃烧的灌木丛中持长矛等待着的人类很容易将这些动物抓走。

采用农业生活方式的第二阶段将涉及狩猎、采集和野生植物种植的结合。这样，人类就能学习并理解种植技术。



在学会了如何种植植物之后，早期的种植者通过寻找前文中介绍过的驯化综合特征，得以将注意力转移到更好植物的选择上。因此，这种转变过程的第三阶段就包括减少对狩猎和采集的依赖，以及更多地依赖于农田生产。这些农田本将包含越来越多的植物和动物驯化品种。奥哈洛二号遗址的证据表明，在10 000年前，只有10%的栽培植物会表现出驯化综合特征。这一比例在8 500年前增长到36%，而在7 500年前增长到了64%。同时越来越明显的是，不同的作物在以不同的速度被驯化，这种驯化速度不是恒定的，而是会被突然的活动打断，而且不同的作物也在以不同的方式被改变。

现在人们认为，尽管中国人早在12 000年前就开始食用水稻，但事实上，他们最早开始驯化的谷物是大约10 000年前的小米。在中国西部的成都平原，小米在4 000年前曾是人类的主食，人们用它来制作面粉、粥和啤酒。来自中国的证据表明，水稻的驯化是一个缓慢的过程。人们仅在一处定居点就检查了大约24 000份植物，其中就包括2 600份水稻小穗。虽然水稻的驯化可能在10 000年就已经开始了，但直到6 900年前，水稻也才占据了栽培植物的8%，而且其中只有27.4%是驯化水稻。到6 600年前时，水稻占据了所有栽培植株的24%，但只有38.8%的水稻被驯化。随着种植面积的扩大，为了容纳农田而发生的环境变化程度也不断扩大。在长江下游地区，人们大约在7 700年前清理了桤木，而通过堤岸对水位进行了150年之久的管控，直到大约7 500前发生了一次灾难性的洪水。

智人进化成依赖农业的物种的最终阶段涉及耕地面积的增加以及农业生活方式的传播。现在人们认为，尽管南瓜、花生和木薯生长在10 000年前的安第斯山脉，但它们最初是在低地的热带森林中发生驯化的。这方面的证据来自对野生和栽培物种的基因组研究。西亚马孙河流域土方工程的证据还表明，这里不仅可能是南瓜、花生和木薯的驯化中心，还可能是辣椒、黄豆、橡胶、烟草和可可的驯化地点。

人们认为，玉米曾经在7 000—5 000年前生长在安第斯山脉。然而，我们不知道这些植物与现代玉米培育的祖先原始墨西哥类蜀黍有何不同。虽然现代小麦和大麦似乎都是在不止一处地点培育成的，也有不止一个原始杂交种或选择，但现代玉米都来自同一个驯化事件，这一事件发生在9 000—6 000年前之间。这种转变令人难以想象。例如，墨西哥类蜀黍的最大谷粒数约为12粒，而一个玉米棒子现在包含多达500甚至更多的玉米粒，这是由于人们不断选择谷粒更多的植物，并且只对其进行种植和杂交。

人们现在认为，最早的独立耕种行为发生在24个以上的地区，其中13个地区的主要作物是谷物。根据经验，增加种子大小和不易碎的种穗是人们最早开始研究的两种驯化综合特征。对于西葫芦来说，人们不仅选择了更大的种子，还选择了较粗的茎和较薄的果皮。

在这二十多个驯化中心出现之后，随着人类的迁徙，农业开始在世界各地散播开来。有时人类迁徙时会带上植物。例如，人们在美国10 000年前的定居点发现了葫芦（*Lagenaria siceraria*）。有人认为葫芦是由古印第安人可能经过亚洲海岸而非横跨大西洋带到美国的。现有良好证据表明，欧洲早期的种植者是移民，教会了土著居民如何耕种。现在还不清楚这些移民来自哪里，但他们携带植物的栽培品种使其远离亲本种，如此阻止了这些品种与亲本发生回交所造成的选择遗传稀释，因而无意中加快了驯化的进程。实际上，他们正在分离基因库。

人类在驯化食用植物的同时，也从植物中提取其他的自然资源。我们在前文中已经提到过橡胶和纤维，但在建造住所并为其供暖时，人们也会利用树木的木材和树叶。然而，植物的另一个主要使用领域是药物。

很难说早期的原始人类是在何时以何种方式开始意识到植物的治疗能力的。同样，洞穴遗址中的碎片给了我们一些线索。人们在伊拉克北部的洞穴中发现了大量的花粉粒，据推测这些花粉粒来自洞穴居民所利用的植物，这些居住者有可能是尼安德特人、智人或者是最近发现的混血人种。人们已经鉴定出，大约90%的花粉粒来自因药用特性而仍在该地区被人类利用的植物物种。这些属性的发现过程仍是人们猜测的主题，但有以下几种可能性。

我们无法排除这些植物特性的发现是简单偶然观察的机缘巧合；当然，人们就是这样发现有毒植物的。观察动物可能使人们对植物产生好奇。獴、马和考拉分别会食用紫锥菊、柳树和茶树。这些植物都已被证明具有益处。各种各样的奇怪想法对人类知识构成做了一小部分贡献。而在知识构成中处于首发位置的必然有形象学说，这一学说的基础在于，如果某种植物的一部分类似于身体的一部分，那么这种植物就应用来治疗该身体部分的疾病。只有用于促进分娩的马兜铃（*Aristolochia*）和用于治疗睾丸癌的北美桃儿七（*Podophyllum*）被认定为这种迷人的 99 危险思想的伪证据。

图19 几个世纪以来，马兜铃一直用于诱导分娩

中药可能是最为广泛的草药使用形式。最古老的印刷版草药志是4 000年前在中国印刷的《本草纲目》[①]。这本著作中推荐的许多注射剂和酊剂至今仍在中国使用，而且仍然有效。科学问题之一就是探索这些配制品中的活性成分，因为配制品通常含有不止一种植物，所以其活性成分可能是多种成分共同作用的综合结果。中国目前正在进行的一项大型项目旨在发现中药中使用的所有6 000种植物的有效成分。

中国人使用麻黄（*Ephedra*）控制支气管哮喘的历史已有3 000多年，而麻黄在当今仍然是儿童咳嗽药中的常见成分。欧洲麻醉师也采用麻黄来预防患者在手术中发生咳嗽——这可是个坏消息。他们也会在患者苏醒过来时使用麻黄，因为麻黄可以模拟肾上腺素的作用，从而使患者感到精神振奋、可以准备回家。应当指出的是，国际奥林匹克委员会和其他机构将麻黄碱及其衍生物视为兴奋剂。黄花蒿（*Artemisia annua*）在疟疾治疗和预防中的应用直接来源于中草药，但其作用方式仍不清楚。多年来，木蓝（*Indigofera*）的提取物一直被用于治疗“坏血”，现在在西方也用于治疗白血病。

在中国，使用银杏（*Ginkgo biloba*）治疗循环系统问题（尤其是与大脑有关的问题）仍有待证实，但作者的母亲发现，在全科医生的处方均无效的情况下，银杏对改善她的手脚血液循环是非常有效的。然而，她的心脏正依靠洋地黄毒苷的非凡能力来调节和加强心跳。19世纪，伯明翰还没有如今这么大规模，而板球运动也还未被创造出来，在埃德巴斯顿村庄工作的威瑟林医生首次证明了毛地黄（*Digitalis purpurea*）、新近的狭叶毛地黄（*Digitalis lanata*）中

① 《本草纲目》成书于16世纪，此处似应为“400年前”，原文应有误。——编注

的地高辛以及随后的洋地黄毒苷的最初用途。他从什罗浦郡一种由20种草药组成的吉卜赛酊剂中偶然得到这个想法，这种酊剂据称对多种疾病有效。经过排除法，毛地黄成了当时的英雄。

随着科学方法的出现，人们对植物提取物的治疗价值进行了更为严格的检验。问题就在于找到可供验证想法的志愿者。在16世纪，如果监狱中的囚犯允许医生在其身上观察各种治疗的影响，那么类似于假释委员会的早期组织便能让囚犯提前获释。提前释放是有保证的——但通常被放在了棺材里。人们便是用这种方法发现了菟葵的毒性。 101

临床试验如今受到了更严格的监控，新治疗方案的寻找也更加系统化。然而，地方性知识的价值仍然不可估量。紫杉醇的发现便是北美原住民与美国国家癌症研究所分享传统医学知识的直接结果。从短叶红豆杉的树皮中提取紫杉醇并在小鼠中证明其治疗卵巢癌的潜力后，一家商业制药公司便开始从事相关研究。他们面临的第一个问题就是获得足够的紫杉醇以进行研究。合成紫杉醇是不可能的，所以唯一的选择就是从树皮中收获这种物质。尽管这帮助了公司，最终也帮助了病人，但对树木却毫无帮助。这种方式永远无法保证紫杉醇的可持续供应。 102

在欧洲，研究人员无法获取短叶红豆杉的树木（因为它们属于美国），所以他们必须找到一种替代的生产系统。最终，人们发现欧洲红豆杉的叶子中含有Bacattin III，这种物质可以转变成具有医疗功效的紫杉醇，随后制成用于治疗乳腺癌的泰索帝。除了表明经过临床证明的新型现代治疗方案仍然来自植物之外，这个故事还提出了另外三个重要的问题。第一，我们应该照顾所有的物种。即使我们现在不会利用它们，但在将来却有可

图20　毛地黄的花

能利用它们。第二，生物资源的可持续开采非常重要。生物学中没有可透支的信用卡。也就是说，人类正在以惊人的速度耗尽我们的储备。据计算，人类在一年中从化石燃料中释放出的二氧化碳需要植物经历300万年的光合作用才能固定形成。第三，这个故事也提出了所有权的问题。美国认为紫杉醇属于他们，其他人都不能利用它。（最后，事实表明美国拥有短叶红豆杉的所有权，而非紫杉醇的专利权。）这三个问题正是构成《生物多样性公约》框架的三个主题。

以生物方式存活是人类挥之不去的重要问题。光合作用是为人类提供日常食物以及多种物质的唯一途径。它有可能增加人类许多主食的每英亩产量。最新的绿色革命是由诺曼·布劳格等人发起的。1970年，布劳格被授予诺贝尔和平奖，以表彰他对世界上营养不良地区的农作物产量的影响。1966—1968年间，印度的小麦产量从1 200万吨增加到1 700万吨。水稻的植物育种计划也带来了类似的产量增加。可悲的是，亚洲的成果从未转移到非洲农业中，这可能是由于非洲缺乏道路、灌溉和种子生产等基础设施。布劳格的准则是“衡量我们工作价值的标准是对种植者土地的影响，而非学术出版物”，但这与英国当前的科学研究基金的准则并非完全一致。

据估计，1960—2000年间，世界上一年中至少有一段时间感到饥饿的人口比例从60%下降到了14%，尽管后者仍然意味着近10亿人口。下一场绿色革命必须进一步提高产量。如果能够减少病虫害的浪费与损失，提高产量是有可能的。种植高粱等耐旱作物可能是另一种选择。一种令人兴奋的可能性是改变水稻的光合作用机制，使其更类似于玉米。（如果各位读者愿意了解更多细节的话，水稻是C3植物而玉米是C4植物，这意味着水稻将二氧化碳转化为含有三个碳原子的分子，而玉米则将二氧化碳转化为含有四个碳原子的分子。）这需要对水稻的遗传基因进行一些巧妙的修改。这种修改似乎在进化过程中发生了45次，但人类对水稻的修改有所不同，有些人害怕人类的食用植物遭到进一步的基因改造。

转基因作物的风险可分为两大类：人畜保护和生态保护。[104]在英国，目前已经有法律程序确保销售食品的食用安全性。生态保护则更加困难，但是可以接受测试和评估。我们或许可以由此总结出三大风险。基因改造的品种是否会逃离种植区域？它是否会与本地物种发生杂交？它是否会对同一种作物的非转基因植物进行异花授粉？如果这些问题的答案都是否定的，那么风险或许是可以接受的。最终缺乏替代品将推动人们接受转基因作物及其带来的食物。



1960年代及以后的绿色革命依赖于高投入的水和肥料。我们知道，磷供应可能会在21世纪中叶逐渐消失，紧随其后的就是无机氮的供应。淡水供应也不是无限的。在为我们的车辆生产生物乙醇和生物柴油的过程中，这也是一个尤其相关的问题。据计算，尽管石油开采和提炼每千瓦时的消耗可能多达190升水，核能每千瓦时消耗多达950升水，但玉米乙醇的生产每千瓦时需要多达867万升水，大豆生物柴油生产每千瓦时需要多达[105]2 790万升水。看来生物燃料会让我们都陷入缺水的境地。

第七章

# 人类对植物的保护

1992年6月5日，在巴西里约热内卢举行的第一次地球峰会上，《生物多样性公约》（CBD）正式启动。迄今为止，绝大多数国家都已经签署并批准了这份公约，它旨在保护生物多样性的组成部分，确保生物多样性的可持续利用，并促进利用获益的公平分享。作为这一里程碑式协议的结果，广义上来说的保护活动大量增加。每个国家都制订了行动计划，并承担起在本地区实施CBD的责任。这种情况的发生得到了这样一个事实的支持，即为保护而筹集的资金中有90%都花在了资金的来源国。其中一个例子就是英国皇家植物园邱园的千年种子库计划，该计划的首要目标是收集和安全储存几乎所有的约1 500种英国本土植物。除了实践活动的增加，标题中提及生物多样性的科学论文数量也发生了爆炸性的增长。

到20世纪末，许多植物学家（尤其是那些在世界各地植物园工作的植物学家）都清楚地认识到，尽管已经存在大量的活动，我们却没有办法知道CBD的愿望是否实现。在2000年4月的一次会议之后，参会者发表了《大加那利岛宣言》以呼吁《全球植物保护战略》（GSPC）的制定，并建议到2010年实现12个以上的目标。已批准CBD的国家对这一提议进行了讨论，并于2002年9月通过了GSPC，预计到2010年实现16个目标。这也许是一个惊喜，因为在此之前从未有人针对一类主要的生物种群提出以目标为导向的策略。其他保护组织对这一策略是否有效很感兴趣，因为这有可能成为未来所有保护工作的范例。到2010年，很显然其中一些目标已经实现，另一些目标将要落空但差距不会太大，还有一些目标将会严重落空。因此，人们在2010年起草了《全球植物保护战略2011—2020》。这显然是基于《全球植物保护战略2002—2010》的成败而设立的；与十年前相比，2010年的植物保护工作进展情况要好得多。



GSPC是一个极好的框架或者说是清单，如果我们要实现CBD中与植物有关的愿景，那么这项战略就列出了我们需要实现的目标。这16个目标涵盖了与植物保护有关的所有活动范畴。它令不同的组织能够做出与其资源和任务相称的贡献；人们认识到，无论是单个组织还是个人都不能使世界上所有的植物物种免于灭绝。为了了解遏制与扭转植物物种减少的任务规模，并认识到这并不是一个无法克服的难题，我们应当先了解GSPC的各个目标。

目标一是编制世界上所有已知植物物种的在线清单。这项工作是由英国皇家植物园邱园和密苏里植物园合作进行的，这两个植物园拥有世界上最全面的植物标本收藏。汇编这份清单所存在的问题是物种的分类存在两种方式。一种方式是根据对某个属中的所有物种（如果属很大的话也可以是某个属的一部分）进行调查的专著进行整理。这项工作通常是由对物种多样性具有全局观点的个人完成的。另一种分类方式就是根据植物群中的国家清查目录。虽然民族自豪感常常成为保护工作强有力的推动力，但众所周知的是，当地植物学家经常通过把小的变种提升到物种的等级来争取夸大所在国家的物种数量，而这些小的变种只是一个物种中正常的多样性变化范围。这种毫无恶意的沙文主义意味着你不能只是把植物群里的所有物种简单累加起来。你也不能仅仅把所有合法描述的物种都列出来，因为大多数物种都不止一次被命名，所以我们还要把同物异名整理出来。

作为创建世界植物群的第一阶段，编制世界清单的工作已经按照植物学分类层级中的科进行了整理，而其中也显示出了一些异常事物。无知的领域似乎毫无规律可循。小家族和大家族一样困难。无论是出于经济原因还是审美乐趣，那些有着悠久栽培历史的科并不比那些只是在最近才引起植物学关注的科更容易被人理解。因此，这项工作揭露出了缺乏专业知识的领域——这在欧洲被称为“分类学障碍”，而在澳大利亚被称为“分类学深渊”！

在整理好物种列表之后，我们需要知道哪些物种正在艰难生存，这就是目标二。在国际自然保护联盟（IUCN）的支持下，各个国家和地区都有汇集濒危物种“红色名录”的悠久传统。在许多国家，这份名录是根据国情而汇集的，但奇怪的是，这份数据也和植物群一样受到了夸大。更长的国家红色名录似乎被视为一种从资助机构撬动更多资金的方式。国家名录还导致在一个国家很常见但在邻国很罕见的物种似乎受到了威胁；误报是很常见的。

现有一份全球名录对所有分类群组中的47 677个物种进

108 行了评估。目前（2009年）的这份名录显示，有17 291个物种（36%）将在未来的50年里面临灭绝的威胁。关于植物的统计数据则更加糟糕，在12 151个已评估的物种中，有71%的物种处于灭绝、野生灭绝、极危、濒危或是低危的状态，仅有12%的物种被列入略需关注[①] 的等级分类。这些数据揭示了一个与数据收集有关的问题，那就是定义。国际自然保护联盟的分类具有明确的定义，并且成为黄金标准，但这需要大量的时间来进行编写，而我们可能没有这么多时间了。

显然，我们首先需要一种启发式方法。人们已经设计出这样一款名为RAMAS红色名录的评估系统，该系统将分类简化为三类：可能受到威胁、可能不受威胁以及可能缺乏数据。或者换句话说就是：受到威胁、没有问题或者不知道。这使得工作人员可以快速输入数据，而无需冗长的计算和全面的细节。除了为问题提供基础更为广泛的粗略估算之外，这款评估系统还将显示数据缺乏的位置，从而显示出哪里需要更多的资源。现在预计所有新的专题论文都将包括对所描述物种的濒危状态的评估。对于木兰和槭树等部分属，研究人员已经在全面的专题修订中对其进行了基于属的评估。目前也存在对于濒危物种数量的其他估计，其物种数量所占比例通常在22%～47%之间。因此，我们有理由认为通常所说的36%的濒危比例不无道理，如果



① 国际自然保护联盟的物种濒危等级系统包括灭绝、野外灭绝、极危、濒危、易危、低危、数据不足和未评估等8个等级，其中低危又分为依赖保护、接近受危、略需关注等3个亚等级。略需关注指该分类单元未达到依赖保护，但其种群数量接近受危类群。——译注

有352 828个开花植物物种，那么在未来的50年里将有127 018个物种面临灭绝的威胁。

我们已经很清楚工作人员正在完成大量的工作，如果只是要达到目标一和目标二，人们还需要完成更多的工作。保护生物学的各个领域都需要进行扎实的科学研究，而这些研究的实施与传播就是目标三。尽管同行评审的期刊终将占有一席之地，但同样真实的是，仍有大量的工作无人报道，而这些研究成果对保护工作者非常有帮助，尤其是那些以发展中国家为研究领域的保护工作者。自2002年以来，已涌现出若干基于网络的资源，基于证据的保护工作均可公布于此。保护项目中反复出现的一种经验就是，每个项目都略有不同。每个物种的恢复项目不尽相同，每个栖息地的恢复项目也有所不同，因为不同地点的物种集聚是不同的。虽然《生物多样性公约》颂扬并保护生物多样性，但正是这种多样性使概括变得困难、规则变得宽松。





因此，公约的前三个目标与了解和记载有关，没有这些，接下来的七个目标就无法实现。目标四是至少保护15%的世界主要生态地区。据估计，目前世界上有11.5%的陆地正受到某种形式的保护。在中国，这一比例是14.7%。保护生态地区的好处是可以保护上一章所提到的生态系统服务。此外，人们通常认为，如果某地具有健康、稳定、多样化的植物群落，那么动物、真菌和其他生物也会定居在此。正是由于这个原因，植物物种的多样性经常被用作所有生物多样性的替代指标。这不无道理，因为大多数生态系统都是通过占主导地位的植物而进行描述的。进一步说，保护大片地区不仅是在保护所记录的物种，也是在保护未知的生物。对于昆虫来说尤其如此，人类迄今可能只描述过

10%的昆虫物种。但这对植物而言并不那么重要，因为人们认为有90%的植物物种至少被命名过一次。

虽然生态系统服务的保护具有不可否认的重要性，但同样不可否认的是，通常情况下植物的多样性，特别是个体物种的分布，在世界各地并不均匀。目标五将解决这一问题，即保护重要110植物地区的75%。这里的目标是指尽量避免所有纬度和各种生境中的植物覆盖率出现差距。众所周知，通常而言，离赤道越远，单位面积上的植物种类数量就越少，但这并不意味着温带地区的重要性和价值就不如热带地区。人们还发现需要进行一些非常复杂的数据分析才能决定将保护区置于何处。马达加斯加的一项研究试图优化2 138种动植物的保护区分配。研究人员

图21　位于牛津郡的帕尔默牧场仅用了三年的时间就恢复为野花草地。自1950年以来，英国的此类草地中有96%都经过耕作

只获得了岛屿面积10%的保护许可，但每一组植物和动物需要的10%略有不同。在菲律宾，另一项研究表明，虽然许多植物在自然保护区得到了充分的保护，但受到威胁的棕榈树大多生长在这些地区之外。

虽然保护“天然”植被地区总是被视为一个好主意，但我们不应忘记的是，地球上25%的土地正处于某种生产体制中。这并不一定是需要投入大量化肥、杀虫剂可能还有水资源的集约式农业系统。它可以是威尔士的山区农地，也可以是伊比利亚半岛的栓皮栎林地。后者是商业支持生产系统的极佳案例，这种生产系统所拥有的相关植物群可令该地区的植物多样性排在全世界的前20位。目标六的目的之一就是确保75%的生产土地是根据保护植物多样性的原则进行管理的。与种植者的田地相比，这个目标在林地更容易实现，然而在东安格利亚，人们经常看到石鸻在集约种植的甜菜地里筑巢。在世界范围内，多达60%的林地将生物多样性保护写入了该地区的管理目标。





所以这三个目标都与以植被为基础的大规模保护有关。然而，许多保护工作都是在物种层面进行的，尤其是因为如果人们捍卫并保护某一物种，那么他们就能看到自己产生影响。以物种为基础的保护工作有缺乏协调的风险，被选中的物种不一定是最需要保护的物种，而可能是最具代表性的物种。例如，资助兰花的保护工作要比资助某种藓类植物的保护工作容易得多。也就是说，大量的保护工作是在物种层面进行的。

多年来，我们的最终目标是对所有物种进行就地保护，如果物种在其栖息地的数量正在减少，那么我们就会进行更多种植以重新引入这一物种。这种做法的失败率远远高于成功率，因为如果物种衰退的威胁或原因没有得到消除，那么这些新植物将与它们的先辈走向同样的结局。唯一可以轻易消除的威胁就是人类的过度采集。这令人们开始考虑迁地保护，即植物在栖息地以外进行生长，例如植物园或者树木园等。结果就是，保护工作两极分化成两个阵营，《全球植物保护战略》的目标七和目标八就能反映出这种分歧。目标七是使75%的受威胁物种得到就地保护，而目标八则是对75%的受威胁物种进行迁地保护，其中10%应保存在物种的起源国。

112

数量正在衰减的物种理应被保存在受管理的自然保护区。这种方法的优点是，植物与真菌、传粉者、扩散媒介以及其他植物之间的所有其他关系能够得以维持。此外，人们普遍认为原位种群中保存有更多的遗传多样性，物种因而能够在适应环境变化的过程中发生进化。这一切可能都是事实，但是植物对于传粉者发生改变的耐受性可能要比我们之前想象的更强，很少有植物会利用扩散媒介，而维持植物种群遗传多样性所需的植物数量也比你想象的要少得多。要保持某一个种群95%的遗传多样性，只需要50～500株无亲缘关系的植物就足够了。就地保护的主要缺点是场地的安全性。在过去，这通常被看作防止人类开发或者栖息地转变的保护。现在可以看到，自然保护区还有另外两种威胁可能更难控制。一种是以后会有越来越多的外来物种入侵，另一种就是气候变化。

气候正在发生变化，这是无可争辩的事实。智人是否是背后因素并因此能对其进行控制，在这里并不重要。世界上的植物以前也经历过气候变化，最近一次就是上个冰河时代末期的恶作剧。植物可能是通过迁移、适应以及利用迄今未开发的特性等一系列综合策略来应对气候变化的。这些策略的相对贡献可能因物种和栖息地的不同而不同。如果迁移十分重要，而我们知道许多植物确实在过去的300万年里做了长距离的迁移，那么城市和农业景观对自然栖息地的入侵可能会使迁移的机会化为泡影。在这一阶段改变人类景观规模的想法是不切实际的，但在静止的自然保护区中保护植物的政策同样存在问题。连接二者的廊道策略是经常提出的一种想法，景观保护也是这样，保护区的位置借此得到协调。 

就地保护的所有优势相对而言都是迁地保护计划的劣势。然而，二者之间存在一种折中的物种恢复计划，其目标是在某处建立起物种的自繁殖种群。为了实现这一目标，人们需要能够提供植物所需要的一切。我们需要了解栖息地需求、传粉生物

图22 濒危植物锐刺非洲铁（*Encephalartos ferox*）的迁地栽培幼苗

学、种子储存需求，还需要有人提供持续、无休止的监测以确保植物的生存。这一策略在西澳大利亚州得到了支持，那里正在实施很多先进的保护措施。通过这样的物种恢复计划，植物物种已从灭绝的边缘恢复过来。

与就地保护相比，迁地保护一直被看作相对较差的一种保护方式，但种子库的出现可能标志着其时机已经成熟。虽然千年种子库计划的首要目标是英国的植物群，但它还有一个次要目标，那就是收集来自世界各地的25 000个物种。这一目标已经实现，而人们又设定下更加雄心勃勃的新目标。在未来气候如此未知的世界里，像千年种子库计划这样的种子库看起来是个好主意。所有批判，例如遗传侵蚀、活性丧失、在战争等随机事件面前的脆弱性，都可以通过合理的采集方案、研究和重复采集来解决。的确，我们在过去20年里对种子的了解在很大程度上是种子库兴起的结果。将来，种子库中储存的种子可能会用于促进植物在自然保护区之间的协助迁移。的确，不是所有的种子都适合脱水和冷冻。也许多达30%的物种都具有所谓的顽拗性种子，我们需要为这类种子找到替代措施。

种子银行并不是一个新概念。它们最初是作为主要农作物的栽培品种基因库而存在的，而世界范围内有许多主要农作物的种子库，例如小麦、水稻、玉米和蔬菜。这些农作物中有许多是一年生植物，因此非常适合储存，也符合《全球植物保护战略》目标九的目的，即主要农作物70%的遗传多样性得到保存和保护。有些农作物并不依靠种子繁殖，所以对于马铃薯和果树等农作物来说，种质圃则成了替代性的选择。在许多国家，业余的园艺家们正在维持早先品种的存活。这些品种可能具有我们将

来需要的独特性状，以传授抗病或抗旱能力。 115

我们在本书中已经提到过外来入侵物种所造成的问题。这是造成生物多样性丧失的五大主要原因之一，这五大原因的首字母可缩写为HIPPO，即栖息地改变、入侵物种、污染、人口增长和过度开发。达尔文的进化论是指通过自然选择使那些更适应其生存环境的生物发生进化，但这一理论经常被错误地简短引述为“适者生存”，而实际上应该是“略胜者和幸运者生存”。达尔文在150年前观察到，世界上并不存在物种极其丰富且资源充分利用，以至于没有外来物种生存空间的地方。他以南非的开普敦为例，那里的物种密度可能比地球上其他任何地方都要高。

有些人担心，最适合在南非生长的生物实际上是在澳大利亚和欧洲进化而来的，反之亦然。外来植物能够胜过本地植物有多个可能的原因。一个是捕食者释放的概念。这是指植物离开在本土对其进行控制的食草动物、害虫和疾病，而在新住所中不受攻击的可能性。无论何种原因，由A国带到B国的植物成为入侵物种，从而对B国的植物群落造成不可逆的损害的概率为千分之一。这种概率或许不低，但有70 000种不同的植物目前正在英国的苗圃中降价出售。这些全功能的基因组对其他植物的潜在危害要远远大于经过基因改造以抵抗除草剂的玉米。

《全球植物保护战略2002—2010》的目标十是为100种最具破坏性的外来物种制定控制措施。这一目标已经实现，并在《全球植物保护战略2011—2020》中被修改为更加明确控制所有国家重要植物地区的外来物种，以及更加明确控制进一步的入侵。在控制外来物种时遇到的一个问题是预测哪些物种会成为入侵物种。物种无法入侵是有原因的：例如土壤的pH值不对、冬天 116

太冷、夏天太干、传粉失败、传播失败等等。然而，不幸的是，对于潜在入侵者并没有行动方案；我们只能成为事后诸葛亮——这方面与经济学非常相似。唯一的控制措施就是预防。我们应该对外来物种关闭边境，就像已经发生在美国、澳大利亚和新西兰的状况一样。

《生物多样性公约》明确表示，我们正在非常贪婪地消耗光合作用的产物。《濒危物种国际贸易公约》（CITES）应该已经阻止了对衰退物种的开发，但目标十一中重申了没有任何植物物种应受到国际贸易的威胁。有时，我们很难知道人们正在交易哪种植物。当某种植物被制成一套遮光帘或是一种草药时，你该如何鉴别这一物种呢？答案可能就在眼前。物种DNA条形码的实践已经在动物中取得了成功。办法是找到一段在同一物种的不同成员之间区别非常少，但在物种之间区别很大的DNA。我们所要做的就是从选定的基因组区域中提取部分DNA并进行测序，将这段序列与数据库中的所有已知物种进行比对，然后就能找到该样品的名称或是被告知新物种的发现。这听起来像是科幻小说，但这几乎就是现实，应用于该领域的手持机器原型正在测试中。下一步就是要创建序列数据库。

虽然目标十一解决的是已濒危物种的保护工作，但目标十二最初旨在通过确保至少30%以植物为原料的产品来自以可持续方式管理的植物来源，以防止其他物种被添加到濒危物种名单上。修订后的目标十二则是，所有野生收获、以植物为原料的产品都必须具有可持续的来源。购买任何物品的任何人都可以为实现这一目标做出贡献。目前已有许多大规模计划令人们能够减少自己对世界生物资源的影响，例如有机食品和林业管



117

理委员会（FSC）计划。与生物学的许多方面一样，《全球植物保护战略》的这一方面也存在灰色地带。在世界上有猩猩居住的地区，关于棕榈油生产的争议就是这样一个问题。剥夺某些人创造收入的能力是需要我们严肃对待的问题，不应轻率地采取行动。

与生物资源丧失相伴随的往往是地方性的本土知识。我们有一系列包含这类信息的惯用语，其中很少是赞美性质的。老妇人的故事或者民间传说就是其中两类。人们常常会发现，这种专业的民族植物学知识并没有被记录下来，因此，当社会被西方知识和西式生活的承诺所取代或吸引时，这些知识的处境就很危险了。如果没有民族植物学家的工作，人们可能永远也不会发现，长期被用作箭头毒物的羊角拗（*Strophanthus*）的汁液中含有被广泛用于治疗心脏疾病的毒毛旋花甙。目标十三希望遏止由于疏忽大意而导致知识损失的无知，但是《全球植物保护战略2002—2010》无法实现之前的目标。

在这一点上，人们遇到了关于贫困和生物保护之间相互关系的长期争论。有些人认为扶贫永远比植物物种的存在更重要，而有人认为如果植物的未来有所需要，那么他们愿意看到饥饿；大家的观点各不相同。和以往一样，现实介于两者之间，许多经济学家现在意识到经济繁荣往往是由健康的生物多样性所支撑的。然而，要说穷人需要对生物的破坏负责，而富人全都关心生物发展，这是很不公平的。世界上植物物种受威胁最严重的地区与繁荣或贫穷无关。

最后三个目标太过开放，我们无法确定是否已经有所影响，但有一些不错的工作实例正在发挥作用。目标十四呼吁在可能



118

图23 羊角拗,可从其中提取毒毛旋花甙来治疗心脏病



的情况下把保护植物遗产的需要列入教育计划。植物不会自己发声,所以它们需要布道者为其说话。植物园、树木园和野生生物信托基金只是提供公共教育计划的三类组织,这些计划旨在强调保护植物及其栖息地的必要性。实现这一目标的另一个方法是将植物保护写入生物学课程的国家教育大纲,也许还要包括经济、地理等课程。

世界各地的植物分布非常不均匀,因而为实现《全球植物保护战略》的目标而努力保护这些植物的人员分布也很不均匀。目标十五旨在纠正这种平衡并确保有足够的经过培训且资源充足的人员人数,以实现战略的各项目标。人们常说没有人会再研究植物学了。如果你的结论只是建立在英国大学课程**名称**的基础上,那么你可能是对的。然而,如果你花时间阅读教学

119

大纲，你会很快发现生物学、环境科学、动植物保护和地理学课程通常包括1960年代和1970年代植物学课程的所有要素。“植物学”这个词有着维多利亚时代牧师住宅和爱德华时代女性的虚假外表，令潜在的学生望而却步，而新课程没有这些特征。然而，植物学在某些国家正处于优势地位。巴西和埃塞俄比亚就是其中两个国家。十年前，培育埃塞俄比亚植物群的项目成员完全由外国人组成。现在这项工作则完全由埃塞俄比亚人进行。

120

在许多国家，千年种子库计划等项目的确在创建合作关系和建设植物保护能力方面非常有效。至关重要的是，参与到植物保护工作的各个方面以及《全球植物保护战略》中的人员正在以一种协调的方式工作。国际、区域、国家和地方各级的网络

图24　许多植物园将培养下一代野外植物学家视为首要任务

是必需的，并且正在根据最后一个目标中的建议而形成。一个新成立的网络将把致力于保护海洋岛屿植物的人们联系起来。这些生物瑰宝通常具有极强的特有分布，因此成了独特的植物群。它们经常面临着共同的威胁。外来物种的入侵在岛屿上尤其常见且具有破坏性。这些网络将国家组织、非政府组织、国际机构、当地人都汇集起来，同时还有业余植物学家和博物学家这些为许多成功的国家战略做出重要贡献的人群。

没有哪个个人或组织能够独自解决物种衰退的问题。现在在《全球植物保护战略》的支持下，我们可以更清楚地看到植物保护工作在哪些方面已经取得了成功，以及下一轮工作必须集中在哪些方面。这项战略已经成为其他领域保护工作的蓝图。

我们现在可以有把握地说，就技术层面而言，人们没有任何理由不能使某个植物物种免于灭绝。在一些国家，《全球植物保护战略》的所有目标几乎都将达成。其中一个国家就是英国，在这里，400年的植物学历史以及实地考察和植物学记录传统表明，如果土著居民投入够多，那么植物和永远依赖植物获取一切的人类都将拥有美好的未来。



121

# 索　引

（条目后的数字为原书页码，见本书边码）

## A



## B



## C

## D



## E

## F



## G



## H

## O



## P



## Q



## R

## S



## T



## U



## V

## W

## X

## Y

## Z

Timothy Walker

# PLANTS

## A Very Short Introduction

# Contents

# List of illustrations

# Chapter 1
# What is a plant?

Plants, like love, are easier to recognize than to define. At the entrance to many areas of outstanding natural beauty in England can be seen a sign that asks visitors to avoid 'damaging trees and plants'. It is fair to ask in what way is a tree not a plant. A plant is often defined simply as a green, immobile organism that is able to feed itself (autotrophic) using photosynthesis. This is a heuristic definition for plants that can be refined if some more characters are added. Sometimes plants are described as organisms with the following combination of features:

1) the possession of chlorophyll and the ability to photosynthesize sugar from water and carbon dioxide;
2) a rigid cell wall made of cellulose;
3) storage of energy as carbohydrate and often as starch;
4) unlimited growth from an area of dividing and differentiating tissue known as a meristem;
5) cells with a relatively large vacuole filled with watery sap.

So trees are clearly plants, and it is not difficult to think of other organisms that are unequivocally plants even though they lack one or more of these characteristics. For example, the orchid *Corallorhiza wisteriana* has the flowers of an orchid, produces tiny seeds typical of the family Orchidaceae, and has the vascular tissue that you find in the majority of land plants. However, what it does not have are green leaves, because this orchid is mycotrophic, meaning that it lives off fungi which themselves derive their energy from decaying material in the forest floor. It is able to do this because of a very intimate relationship with a fungus, a characteristic found to varying degrees throughout the orchid family. In a similar vein, *Lathraea clandestina*, which can be seen growing on the banks of the River Cherwell in Oxford, has flowers reminiscent of a foxglove, yet it too has neither shoots nor leaves. Its flowers emerge directly from the soil because this plant has roots that are able to infiltrate the roots of willow trees and divert the nutritious contents of their vascular tissue. Both of these plant species have lost the ability to photosynthesize, but they are still plants because they share many, many other features with those plants which do still photosynthesize.



The problem with the definitions above is that they are too limited, because they do not take into account some of the algae that live in water. In order to arrive at a sensible and unambiguous definition for plants, we need to consider how we classify biological organisms. Similar individuals are grouped together into a species. Similar species are then grouped into a genus. Similar genera are grouped together into a family; and similar families are grouped into an order; similar orders into a class; similar classes into phylum; and similar phyla into a kingdom. Each of the groups in this hierarchy can be referred to as a taxon, and the study of groups is known as taxonomy. Prior to the 19th century, taxonomists tried to create a *natural* classification that revealed the plan of the Creator. Since the 19th century, biologists have questioned whether species can change and evolve by retaining those changes and passing them on to their offspring.

A great deal of work is currently being carried out to build the 'tree of life' (or phylogeny) that shows how all living organisms are related to each other. This work received its kickstart in 1859 with the publication of Darwin's *On the Origin of Species*, and it is still ongoing. An evolutionary tree is the only illustration in

1. ***Orobanche flava*** **is one of many parasitic plants that do not photosynthesize but which steal from other plants**

*The Origin*, and Chapter 13 of the first edition is still an eloquent introduction to taxonomy. Darwin talks about the possibility of building a *natural* classification, but now natural means revealing the course of evolution and not the mind of God. Classifications now are based on what Darwin called *commonality of descent*. All the members of a taxon must share a common ancestor, and the group must contain all of the descendants of that ancestor. If these criteria are fulfilled, then the group is said to be monophyletic. Monophyletic groups occur at every rank in the classification from species to kingdom.

If we see the evolution of species over the past 3,800 million years as a branching tree, then plants are one set of the branches on the tree of life, and this set of branches is all connected back to one crutch. The arguments start when you try to decide which crutch marks the start of plants. It is worth saying at this point that fungi are definitely *not* plants. Fungi are in fact on the branch next to animals on the tree of life. Despite this, mycologists (who study fungi) do tend to be grouped with botanists rather than zoologists in university departments.



## The original plants

At the heart of any definition of plants is the ability to photosynthesize. Unfortunately, there are organisms that photosynthesize but which cannot be considered by anyone to be plants. In particular, there are the photosynthetic cyanobacteria.

It is currently believed that life has evolved just once and that this happened about 3,800 million years ago. At that time, the world as an environment for biology was very different. There was no protective ozone layer to absorb the harmful ultraviolet light from the Sun. Furthermore, the atmosphere contained a great deal of carbon dioxide but very little oxygen.

The first living organisms were simple compared to the majority of plants that we see around us today. For a start, they were unicellular. They were prokaryotes. There are many prokaryotic organisms still extant in two big groups: the archaea and the bacteria. (The other major group of organisms are the eukaryotes, that is, plants, animals, fungi.) Fossilized prokaryotes have been found in rocks dated at nearly 3,500 million years old. The fossils of these early bacteria are grouped in structures that look the same as the stromatolites that can be seen in several places around the world today.

A stromatolite is a cushion-shaped rock that is found on the edges of warm shallow lakes, most commonly salt-water lakes, and they are (very simply) laminated accumulations of microbes. The colonies of the unicellular cyanobacteria live in a film of mucus. Calcium carbonate builds up on the mucus and the cyanobacteria migrate to the surface and a new layer of mucus is formed. These alternating layers are then fossilized and the bacteria enclosed in the rocks. So it was clear to see that prokaryotic life had evolved perhaps as early as 3,800 million years ago, but it was not so easy to determine how these early living entities found the energy to live. Some may have synthesized enzymes to break down minerals, but this was slow. There is now compelling evidence that the cyanobacteria in these fossil stromatolites were able to capture the energy of the Sun and use it to synthesize molecules containing carbon derived from the abundant carbon dioxide in the atmosphere. This evidence is based around the fact that the enzyme that drives the capture of carbon from carbon dioxide preferentially fixes one carbon isotope ($^{12}C$) over the other that is also present in the atmosphere ($^{13}C$). So if carbon compounds contain the two isotopes in different proportions from those in the atmosphere, then the compounds were the product of photosynthesis. Carbon compounds have been found in rocks in Greenland that have the carbon isotope ratio produced by photosynthesis.

Photosynthetic organisms, with which we are familiar, use water as a source of electrons. The oxygen in the water is then released into the atmosphere as gas. It is thought that the first photosynthetic cyanobacteria may have used hydrogen sulphide ($H_2S$) rather than water ($H_2O$). It is currently believed that by 2,200 million years ago, cyanobacteria were generating large amounts of oxygen and that this was accumulating in the atmosphere. This may seem like a small point, but the fact that cyanobacteria began using water as a supply of electrons led eventually to the levels of oxygen in the atmosphere that made aerobic respiration possible and the majority of biology as we know it. The generation of oxygen had another effect, namely the formation of the layer of ozone in the upper atmosphere, whose absence has already been noted and whose protective function is so important for biology. Prior to this, the mucus in the stromatolites may have helped to protect the cyanobacteria. Living in water would also have afforded some protection.



So to recap, we see that by 2,000 million years ago, there was a large population of prokaryotic cyanobacteria that was generating oxygen by photosynthesis, but there was still nothing that we could describe as a plant. The evolution of plants required an event that must have happened but for which we do not have a complete cast list. This event was the formation of the first eukaryotic cell. Eukaryotes cells are more organized internally than prokaryotes. They have organelles enclosed by membranes such as the nucleus and mitochondria and, in the case of plants, chloroplasts. Organelles are just small 'organs' found inside cells which perform specific functions within the cell.

It is believed that 2,700 million years ago, an unidentified unicellular prokaryotic organism engulfed another but did not break it down. The engulfed cell retained its membrane and gave up some, but not all, of its genes to be included in the nucleus of the host cell. This engulfing followed by 'cooperation' is known as endosymbiosis. This early eukaryotic organism (known as the proto-eukaryotic cell) lived by metabolizing photosynthetic products from free-living cyanobacteria. The evidence for this endosymbiosis is simple: the organelles have two membranes – their own and one from the host that engulfed them. The evidence for the timing of the first endosymbiosis is equally simple. One of the unique features of all eukaryotes is the production of sterols. When a eukaryote dies and breaks down, the sterols are converted into steranes, and these persist in rocks for a very long time. Rocks 2,700 million years old contain steranes, and so there is a trace of dead eukaryotes but no fossils of intact organisms.

Many years passed and the diversity of eukaryotic organisms increased, resulting in evolutionary lineages that produced many other species (both extant and extinct) but not plants. However, having recruited one type of prokaryotic organism, the proto-eukaryotic cell recruited another, and this time it was a photosynthetic cyanobacterium. As before, the incoming organism became an organelle, and some, but not all, of its genes were transferred to the nucleus of the host cell. Again, as before, the organelle, known now as a chloroplast, has a double membrane.

The oldest fossil evidence of the structure of a probable eukaryote is in rocks 2,100 million years old. The organism, named *Grypania spiralis*, has no extant descendants. It looks a bit like an algae, and so it is believed (or perhaps hoped!) that it was photosynthetic. At 2 millimetres in diameter, it is big enough to be an ancestor of some of today's algae, but it cannot be proved to be this ancestor. The first undisputed fossil of a photosynthetic eukaryote that can be placed in an extant taxon has been found in rocks 1,200 million years old. The organism, *Bangiomorpha pubescens*, is a red algae and is thus named because it resembles the extant red algae *Bangia atropurpurea*. In addition to looking similar, these two species also share a habitat, the margin of land and water.

*Bangiomorpha* is significant for another reason: it is currently the earliest example of a multicellular eukaryotic species that not only has cells with specific functions, but also one of the functions of these specialist cells is to indulge in sexual reproduction. Multicellularity is one of those important biological events that has evolved more than once on different branches of the tree of life. The most recent common ancestor of plants and animals was unicellular and yet both are now dominated by multicellular organisms.

The fossils of *Bangiomorpha* are so well formed that it has been possible to reconstruct its life cycle and it is similar to some of those found in the red algae. The spores germinate and grow into the multicellular body of the plant. The spores contain only one set of chromosomes (that is, they are haploid), so the algae plant is haploid. At the base of the plant is a holdfast that fixes the plant tightly to a rock. At the top, the plant becomes flattened, and as it grows upwards it is able to capture more light. Some of the cells in this thallus differentiated into haploid gametes that are a prerequisite for sexual reproduction.



So at some time between 2,100 and 1,200 million years ago, the first photosynthetic eukaryotic organisms emerged. These were the first plants, and every subsequent organism on this branch of the tree of life is a plant. The two endosymbiotic events that define this branch happened once and are known together as the primary endosymbiosis. The evidence for this has been derived from the analysis of DNA sequences. This technique, available from 1993, has been very important in unravelling previously tangled problems of evolution.

So plants as described in this book are monophyletic. Intriguingly, it is becoming clear that there has also been a secondary endosymbiosis in which some of the true plants have been incorporated into non-plant organisms and the resulting organisms are also not plants. Perhaps the most familiar of these

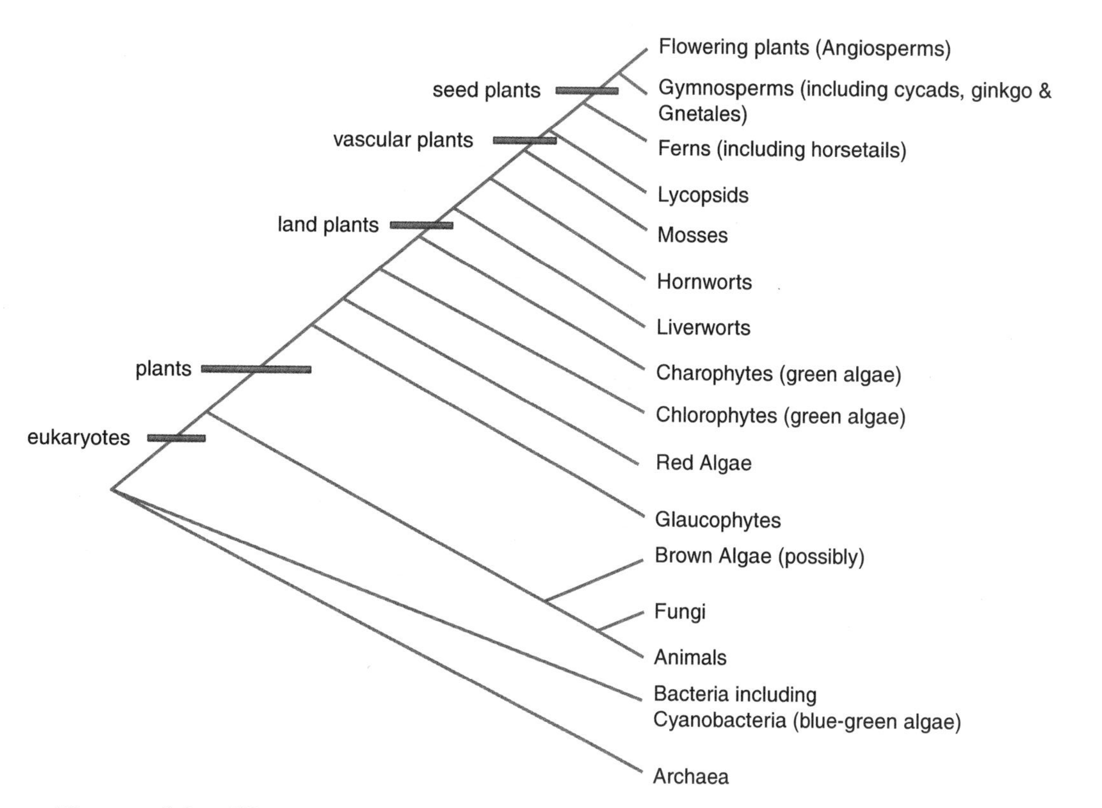
Flowering plants (Angiosperms)
Gymnosperms (including cycads, ginkgo & Gnetales)
Ferns (including horsetails)
Lycopsids
Mosses
Hornworts
Liverworts
Charophytes (green algae)
Chlorophytes (green algae)
Red Algae
Glaucophytes
Brown Algae (possibly)
Fungi
Animals
Bacteria including Cyanobacteria (blue-green algae)
Archaea
seed plants
vascular plants
land plants
plants
eukaryotes

2. The tree of plant life

are the brown algae such as kelp. So if you visit a beach and start looking at the 'seaweeds' left behind by the tide, the green and red ones are plants and the brown ones are animals.

## The oldest extant plants

The lowest side branch on the plant branch of the tree of life consists of a very old (at least 1,200 million years), very small (13 species) group of tiny (microscopic) freshwater algae in three genera – *Glaucocystis, Cyanophora, and Gloeochaete* – known collectively as the glaucophytes. The lowest branch on a phylogenic tree contains the organisms that are thought to have changed the least of any others on the tree. This does not mean that they look exactly like the original ancestors. The evidence that supports their basal position is two-fold. Firstly, in addition to the two membranes around the chloroplasts, there is a peptidoglycan layer. This is similar to the envelope found around bacteria, and the chloroplasts are sometimes referred to as cyanelles to distinguish them from those found in the rest of plants. This peptidoglycan layer is not found in any other plants, the inference being that it has been lost early in the evolution of plants. Secondly, they have pigments called phycobilins in their plastids. Plastids are organelles found in plants where important chemicals are manufactured or stored. These pigments are only found in cyanobacteria, the glaucophytes, and the red algae, which are the next side branch on the plant branch of the tree of life. In all three groups, these pigments are bound into phycobilisomes. In addition to the phycobilins, they have chlorophyll a.

This small group is of interest to evolutionary botanists because it is thought that they are the closest extant species to the original endosymbionts. Some species can move and some cannot, while some have cellulose cell walls and some do not. Sexual reproduction is unknown in this group of plants. The evidence from DNA sequences confirms the morphological evidence, and so clearly these little plants are strong contenders for the title of the earliest diverging branch on the plant tree of life. The next side branch up is the red algae. It is worth reiterating that the term 'algae' is a vernacular term that is applied to a group of organisms, some of which are plants (red algae and green algae) and some of which are not plants (blue-green and brown algae).

## Red algae

There are many species on this branch: somewhere between 5,000 and 6,000, and perhaps as many as 10,000, of which just a small percentage live in fresh water. The red algae share some characteristics with the glaucophytes that have subsequently been lost and so do not appear in the rest of the plants. These characters are phycobilin pigments and phycobilisomes and having just chlorophyll a. These pigments, other than the chlorophyll, give these plants their distinctive red colour. Red algae have other features in common such as storing energy as glycogen (or floridean starch). Glycogen is a large molecule with lots of glucose in a chain off which come side chains of molecules. Some species secrete calcium carbonate and these are important in the construction of coral reefs, hence why these are known as coralline algae. Red algae all have a double cell wall. The outer layer is economically important because it can be made into agar, which has many uses including in cooking. The internal wall is made in part of cellulose, like most plants.

As one would expect, in a taxon as species-rich as the red algae, there is a lot of diversity, but there are common patterns to their life histories. However, that pattern is complicated and very different from the life cycle of mammals with which we are most familiar. This familiarity colours our preconceptions and assumptions about plant reproduction, and it can create a barrier that makes understanding plant life histories much more difficult than it need be.

The first false assumption is that *every* free living organism needs two complete sets of chromosomes just because we do. This is not true, and we have already seen a fossil species (*Bangiomorpha*) which spent much of its life in the haploid state with one of each chromosome, rather than in the diploid state in which it would have had two sets. It is reasonable to ask if there are any advantages or disadvantages attached to being haploid or diploid. A fact of life as a result of being haploid is that any deleterious mutation will be expressed and the organism may perish as a result. So it may appear that having two copies of each gene is better because the harm of a deleterious mutation can be overcome by the good copy, or perhaps the combination of two versions of a gene might be better than just one version. However, this is a twin-edged sword because it means that diploid cells can build up many potentially crippling mutations. Given that there appears to be an evolutionary trend towards diploidy in all the major taxa in the tree of life, it appears that this is, however, a more stable evolutionary strategy. Despite this, it cannot be denied that being haploid has not prevented many taxa from surviving successfully for hundreds of millions of years. Among these taxa are red algae, green algae, mosses, liverworts, and ferns.



The second false assumption that we make is that there will be a group of cells (the germ line) that from a very early stage in the life of an organism will be responsible for making the gametes – the sperm and eggs. This is true in mammals and many other animals but not in plants. In general, the development of plants and the differentiation of their cells into different structures is very flexible and certainly not determined early in the life of the embryo. This can be seen clearly when a gardener roots a cutting, or when a biennial plant like a foxglove changes from vegetative growth and starts to produce flowers. Plants do not have a group of germ cells. Instead, they have a distinct and finite stage in their life histories when a haploid plant produces sperm and/or eggs. Because the plant is already haploid, there is no need to halve the number of chromosomes before the gametes are formed by differentiation and mitosis (cell division whereby the chromosome number remains the same and two identical cells are produced). The haploid stage in the life history produces gametes, and so is very sensibly known as the gametophyte, '-phyte' being derived from the Ancient Greek word φυτόν, meaning plant.

A third false assumption (that is clearly wrong given the previous paragraph) is that diploid plants will produce haploid gametes by the process of meiosis (cell division whereby the number of chromosomes is halved). Instead, the diploid plants produce haploid spores by meiosis. Unsurprisingly, the diploid stage in the life history that produces spores is known as the sporophyte. These haploid spores grow into the haploid gametophyte, and the life history repeats itself. Plant life histories therefore consist of alternating stages or generations: between a haploid generation and a sporophyte generation. Plants are often described as demonstrating an alternation of generations.



Red algae (or their ancestors) were the first plants to display sexual reproduction, and so it is worth describing it here as it makes what comes later easier to understand. What follows is a description of the life history of *Polysiphonia lanosa*. This is a red algae that may be familiar to anyone who has spent time rock-pooling in the intertidal zone of the coast of Britain and Ireland. This species of red algae grows on the outside of *Ascophyllum nodosum* (a brown algae and so not a plant) which is very common all the way up the west coast of Europe and the north-east coast of North America. The *Polysiphonia* is probably an epiphyte on the *Ascophyllum*, but some people have recorded that the former actually penetrates the latter, and this would make it parasitic. The *Polysiphonia* plants look like cheerleaders' pompoms on the surface of the *Ascophyllum*.

These plants of *Polysiphonia* may be male or female gametophytes, but they look very similar. The male plants

**3. Brown, red, and green algae are often washed up on the seashore together, but only the red and green algae are plants**

produce male gametes. These are known as spermatia rather than sperm because they do not have a flagellum, or tail, for propulsion. These are released into the water around the plants and the hope is that they will find a female gametophyte. The female gametophyte produces (but does not release) a female gamete. This is a cell that is retained inside a structure known as the carpogonium which consists of the female gamete and a trichogyne. This trichogyne is a protuberance whose function is to catch a passing spermatium. Once caught, the spermatium will donate its nucleus to that of the female gamete, and a diploid cell is formed known as a zygote.

The zygote develops into a diploid structure that remains enclosed in, and therefore completely parasitic on, the haploid female gametophyte. The diploid structure is known as the carposporophyte, from which we can deduce that this pustule produces diploid carpospores. These carpospores are released into the water and drift around hoping to land on a suitable substrate, which in the case of *Polysiphonia* is an *Ascophyllum* frond. These diploid carpospores germinate and grow into diploid tetrasporophytes. Rather unhelpfully, these tetrasporophytes look very similar to the gametophytes despite the former being diploid and the latter haploid. When a species has haploid and diploid generations that look similar, it is said to be isomorphic. When the two generations look different, they are said to be heteromorphic. Some red algae are isomorphic and others are heteromorphic.



When mature, tetrasporangia form on the surface of the branches of the tetrasporophyte. The tetrasporangia are the sites of the production of haploid tetraspores, so called because they are formed in tetrads (or 4s). This means that they are joined together in a triangular pyramid with each spore attached to each of the other three. These haploid spores are produced by a diploid plant and therefore they are the result of meiotic cell division, whereby the number of chromosomes is halved. The tetrad of spores is released, and hopefully it will land on a frond of *Ascophyllum*. One life history of *Polysiphonia lanosa* is now complete.

The internal structures of red algae are as variable as their life cycles, but *Polysiphonia lanosa* can be used as an example. Each branch has a central axis of elongated can-shaped cells that are joined end to end. These cells are joined by pit connections that form during the process of cell division. Associated with these connections are pit plugs, which are able to seal the connection should one of the cells die. Intercellular connections are an important component of multicellularity. Around this central axis of cells is a layer of periaxial, or pericentral, cells. These are the same length as each cell in the central core and aligned with the cells therein, thus making the branches of the plants look like a series of repeating units. There may be a further layer of cells, known as the cortical cells, around the periaxial layer.

## Green algae

Green algae belong to a much larger and very diverse group of plants called simply the green plants. It is currently believed that they have all evolved from one common ancestor. They share a number of features, including both forms of chlorophyll – a and b – and they have a cell wall made of cellulose. The majority of the green plants are the land plants; the rest are the green algae.



In the same way that 'algae' is a vernacular term that is applied to a group of organisms, some of which are plants (red algae and green algae) and some of which are not plants (brown algae), the term 'green algae' refers to two different groups that do not share one unique common ancestor. The green algae are two branches on the evolutionary tree: the chlorophyte algae and the charophyte algae. The evolution of these plants and their relationships is not yet fully understood; in fact, it is a bit of a mystery because the land plants have received much more attention than the algae in the past twenty years. However, it appears that one branch is very much less species-rich than the other: the charophytes, including the Charales. This smaller group is arguably more important in that it is the sister group of the land plants that now dominate the world.

Rather than try to give a comprehensive survey of the green algae, a few species will be described in detail to illustrate the diversity found in this group. Green algae are found in fresh water *and* sea water. Many species are unicellular but some are filamentous, while others spend some of their time as single-celled organisms that then form a multicellular colony in which some cell differentiation occurs. *Volvox carteri* is a good example of this latter behaviour. Some algae form symbiotic relationships with fungi known as lichens. While the algae can live without the fungus, the reverse is not true, and so lichens are known by the fungus's name not the algae's. Furthermore, one species of algae can form a symbiosis with many different fungi. (Lichens are *not* plants.)

The first group of green algae many people encountered was *Chlamydomonas* because it grew in the school pond. This unicellular genus, along with the *Volvox* and *Ulva* described later, belong to the chlorophytes. *Chlamydomonas* has two flagella and has been used extensively as a 'model organism' in the movement of flagella. The adult plant is haploid, with just one set of chromosomes. This cell can reproduce asexually by simply dividing mitotically. Before these new adults are produced, the *Chlamydomonas* lose their flagella and group together. The cells then divide but in an uncoordinated fashion, producing new unicellular organisms.

However, the adult *Chlamydomonas* can differentiate into a gamete. In some species, the male and female gametes are the same size (isogamous), whereas in others the females are larger (oogamous). The two gametes fuse in the water and form a diploid zygote that is encased in a thicker wall that can protect the young zygote. This zygospore can withstand harsh conditions, but when it is feeling that conditions are correct, it will divide meiotically and four new adult plants are released from each zygospore and the life history is completed.

*Chlamydomonas* is in the family Chlamydomonaceae which is in the same group of families (or order) as the next genus we shall look at – *Volvox*. *Volvox*, like *Chlamydomonas*, has been investigated in depth because it is capable of existing as a unicellular plant or in a spherical colony of several thousand cells. *Volvox carteri* is thus a useful subject if you want to study the evolution of multicellularity, which has evolved many times in plants, let alone independently in plants and animals.

So the unicellular *Volvox* plants come together to form a colony whereby several thousand individuals become embedded around the outside of a gelatinous ball of glycoprotein (known as a coenobium) that is up to 1 millimetre in diameter. This colony works as a collective, with coordinated flagella beatings that can move the ball towards the light. Sometimes connecting strands of cytoplasm can be seen connecting the cells. The cells perceive the light through eyespots which are more common on one side of the colony than on the other, thereby giving the ball a front and back, or anterior and posterior pole. This coordinated group of individuals then becomes a truly multicellular organism when some cells begin to divide asymmetrically to give one small and one larger cell. The two new types of cell are incapable of an independent existence. The small cells are somatic cells whose function is to propel the colony with their flagella. The larger cells, known as gonidia, accumulate at the posterior pole, where they divide and give rise to daughter colonies. These young colonies are initially retained inside the coenobium, with their flagella orientated inwards, but when the parent colony ruptures to release the offspring, the cells reorientate and the flagella are on the outside of the sphere. Vegetative reproduction is complete. Bizarrely, sometimes granddaughter colonies form inside the daughter colonies before they are released by the mother colony.

Sexual reproduction of *Volvox* is also different from that of *Chlamydomonas*. Some species have colonies that produce both sperm and eggs (monoicous), while in other species the colony will only produce sperm or eggs (dioicous). (It should be noted that this is different from monoecious and dioecious which is used when diploid plants are male or female as opposed to bisexual.) At the commencement of sexual reproduction, some of the generative cells in the colony will *either* begin to produce sperm which are released *or* to develop into egg cells which are not. The sperm are produced in sperm packets, which are simply bags of sperm that are released from the parental colony. There is some evidence that these packets release a pheromone to make other colonies sexually active. When the sperm finds an egg and successfully fertilizes it, a thick-walled spore results that contains the zygote. This diploid spore, known as the meiospore, is capable of withstanding harsh conditions, but in the correct circumstances will undergo meiosis and release haploid offspring.

One of the more familiar green algae to anyone who has been rock-pooling along the shores of temperate regions of the world is sea lettuce, or *Ulva lactuca*. This often scruffy-looking plant is found attached to rocks by a round holdfast or drifting freely. The rest of the plant is up to the size of a dinner plate and is a thallus of just two layers of cells thick, making it very flimsy, and so it regularly gets torn by the action of the waves on the rocks. However, this is not a problem as it lives in water and the plant is buoyant and supported by the water. The cells in each layer are arranged randomly and any one of them can divide. This means that there is no equivalent of the meristems that we find in flowering plants. The individual cells are not interconnected in any way, making this little more than a colony in some people's minds. This is an over-simplification as the plants are in fact more organized than that, in that they are like herbaceous perennial land plants because they can regrow from the holdfast each spring or if the thallus breaks off. The pieces of thallus that break off have been found to form new plants in laboratory conditions, but it is not known if this happens in the wild.

**4. Sea lettuce is one of the most common green algae on English beaches**



*Ulva*, like all sexually active plants, has a life history that includes an alternation between a haploid generation and a diploid generation. The twist in this part of the tree of life is that the haploid and diploid generations look the same and so are described as isomorphic. The haploid plants produce gametes. To do this, cells around the margin of the gametophyte thallus divide and differentiate into biflagellated sperm or eggs. These are similar morphologically except that the females are slightly larger. Both male and female gametes are able to photosynthesize and to swim towards a light source (positively phototaxic). This means that the gametes swim up to the surface of the water. It is believed that the flagella are more than just a means of propulsion; they are implicated not only in sexual identity but also as facilitators of adhesion once a gamete of the other sex has been located. The gametes of both sexes have eyespots at the base of their flagella. The gametes differ in that the female eggs have 5,500 particles in the outer membrane of the eyespot chloroplast while the male gametes have 4,900. Having found the gamete of their dreams, the now quadriflagellate zygote is negatively phototaxic, meaning that it swims away from the water's surface to the rocks at the bottom of the water, where they can grow a holdfast and then a new thallus, only this thallus and holdfast are diploid.

When mature, the diploid thallus of the sporophyte produces haploid spores from the margin. These are the result of meiotic division that halves the chromosome number. These zoospores (like the zygotes) are both quadriflagellate and negatively phototaxic. Also like the zygote, the eyespot membrane has 11,300 particles, and this is thought by some to have a role to play in phototaxis. The zoospore will then grow into a holdfast and a male or female gametophyte. *Ulva* is collected from beaches in Scotland and eaten in soups and salads, and in Japan it is used in some sushi dishes.

The charophytes are a much smaller group than the chlorophytes in terms of the number of species. One may be familiar from school days. *Spirogyra* is a filamentous algae found in freshwater pools and ponds. It normally lives below water, but in warmer weather the rapid growth rate and lots of oxygenic photosynthesis results in a frothy, slimy mass of tangled filaments rising out of the water. At any time of the year, it is easily identified by the fact that the chloroplasts are arranged in a pattern that resembles a stretched spring. The cylindrical cells are joined end to end and individual filaments may be many centimetres long. The cell wall has two layers: an outer coat of cellulose and an inner wall of pectin. The filaments can break, but this is essentially asexual reproduction as the two halves can each grow into a new plant.

Each cell of an adult plant of *Spirogyra* is haploid, so it is a gametophyte. Sexual reproduction is simple and comes in two scenarios. In the first, two different filaments come to lie alongside each other. Tubes grow out from cells in each filament and fuse at their tips to create a conjugation canal between two cells, one from each filament. The contents of the male cell migrates into the female cell, the nuclei fuse, and a diploid zygote is formed and released as a zoospore. This is known as scalariform conjugation, like a ladder. The other scenario is known as lateral conjugation. This is when a filament forms conjugation tubes between adjacent cells in the same filament. This is followed by the migration of the male contents into the female cell and the formation and release of the diploid zoospore as happens in scalariform conjugation. This spore then divides meiotically to give four haploid cells, from which new gametophyte filaments form.

The final example for this chapter and the second charophyte is *Chara* itself. This is a multicellular plant that is found in freshwater pools in temperate regions of the northern hemisphere. The plants look similar in general terms to other water plants such as *Ceratophyllum* (see Chapter 5) and to some land plants such as horsetails or goosegrass, yet they are closely related to neither. The plant consists of a central filamentous stem from which whorls of branches are produced at regularly spaced nodes. The plants may be found floating freely, but they do grow into the mud at the bottom of ponds with the production of rhizomes. The cells at the apex of the stem divide, with the upper daughter cell retaining the function of apical cell. The apical cell in a plant is simply the cell at a tip. The lower daughter cell develops into either a nodal or inter-nodal cell. It is from the nodal cell that the whorl of branches grows. The branches are either short and of determinate growth, or long and of indeterminate growth.

This plant is the haploid gametophyte. The plant produces motile sperm that are released into the water. The female gametes are not released but retained in structures on the gametophyte. The diploid zygote that is formed could, and perhaps through our eyes should, grow into a diploid sporophyte, as happens in *Ulva*. However, *Chara* is not us (that is, not human) and the zygote goes straight into meiosis to produce four haploid spores. These spores drift away and develop first into a filamentous protonema and then into more haploid adult gametophytes.

So the green algae consist of two separate groups: the chlorophytes and the charophytes. These are not one monophyletic group sharing one unique ancestor; rather, they are adjacent branches on this limb of the tree of life. There is one more big group left on this limb and it shares a unique common ancestor with the charophytes. This big group is very big. It consists of about 400,000 species. This group of plants is very familiar to us because they are the land plants. These plants can be traced back to one unique, unidentified ancestor that had accumulated a new combination of traits that enabled it to survive for most of the time out of water. This ancestor was the first land plant, and without it there would be no terrestrial ecosystems as we know them, and there would certainly be no *Homo sapiens* and no Oxford University Press and no *Very Short Introduction* to plants. The next chapter is about the land plants.

# Chapter 2
# Living on dry land

About 470 million years ago, a plant survived for more than one generation out of the sea water where the green algae had lived for many years. This is one of the major events in the natural history of the Earth and is often referred to as the invasion of the land. However, it could be just as easily (and perhaps more accurately) described as the invasion of the air, since many of the plants in the water had already evolved means of attaching themselves to the bed rock. So what were the challenges facing plants wanting to leave the familiar comfort of the sea?

Firstly, there was desiccation resulting from the wind experienced when living in the air. When living in the sea, there was water all around that would enter the plants' cells by osmosis. It could be said that keeping water out is more of a problem than keeping it in if you live in the sea. When the first land plants were stranded out of the water, they had not only to reduce loss of water but also to find a means for taking up water to replace the inevitable loss. So hand in hand with the problem of taking up water, there is the problem of holding onto the water. If we look at how some of the older lineages of land plants, such as mosses, have solved this problem, we find that rather than preventing desiccation, they simply tolerate it and then rehydrate when it rains. It should come as no surprise that 10% of the moss species found in Europe are found in Britain, and half of European liverwort species are found in Ireland, because both countries are notorious for rainy weather. So the first land plants possibly led a rather Jekyll and Hyde lifestyle, absorbing water and nutrients when it was wet and shutting down when it was dry.

Secondly, and connected to the first problem, the first land plants had to find a way to allow air (containing carbon dioxide and oxygen) into their photosynthetic structures. For plants living in water, these two substances are dissolved in the water and sufficiently available. Any openings in the leaves for the ingress of gases will also be an opening for the exit of water, which is now very precious.

Thirdly, and related to the previous problems, there was the problem of finding the other raw materials that a plant requires such as nitrogen, phosphorous, potassium, calcium, iron, sulphur, magnesium, silicon, chlorine, boron, zinc, copper, sodium, molybdenum, and selenium. It must be appreciated that the land these plants were colonizing was bare rock. While there may have been deposits of sand and other fine particles as well as larger pieces of rock, there was probably no soil as we now know it. This is because a vital component of today's soils is the organic matter that is the remains of dead organisms. There are also the other living organisms themselves such as bacteria, fungi, and animals. Before there was soil, there was no evolutionary pressure to develop a complex root system to take up nutrients from the soil. There was, however, the need for an anchor that would prevent the plant from being blown or washed from its place on the rocks. The algae have a holdfast but this structure does not seem to have made its way onto dry land.

Fourthly, there was the problem of finding new places to live. Although the plant wants to be attached securely, it still wants to be able to explore the area for potentially better positions and to escape from disasters. This required the emergence of protection for the haploid spores. It turns out that this may have already been in place in the form of sporopollenin. This is a very complex and extremely durable substance found on the outside of the spores (and pollen) of land plants and in the walls of spores of a few green algae such as *Chlorella*. For people studying sporopollenin, its durability is a problem as it makes it very unwilling to reveal its structure and chemical composition, so it is impossible, at present, to be sure that the green algae sporopollenin is exactly the same as that found in land plants. It appears, therefore, that there was a pre-adaptation to solve this problem. Pre-adaptations are not uncommon and are probably a requirement for large advances or multiple innovations in evolution that would otherwise require the simultaneous emergence of several novel structures.

In this instance, one person's problem turns out to be a gift for other scientists, as the durability of sporopollenin enables evolutionary botanists and ecologists to track the distribution of plants by the deposits of pollen in soil cores and even ancient rocks. Some of the earliest evidence for the existence of plants on the land is spores showing liverwort characteristics that have been found in rocks dated at 475 million years old. These spores are 50 million years older than the first fossils of intact plants (or even fragments of plants).

The fifth problem is the delicate matter of the sperm finding an egg. The plants that live in water release one gamete (or occasionally both) into the surrounding water and it (or they) swims off to find a friend. For some botanists, this is seen as a very big problem for land plants, but this may be an exaggeration. The argument suggests that if you need water for your sperm to swim through on its way to the egg of its dreams, then the parent plants are going to be restricted to wet places. There are still some plants, for example mosses, for which water is required to facilitate the bringing together of gametes. Simple observations of where these plants grow show that they are simply restricted to places that are *seasonally* wet. Mosses are often found on the tops of walls, a habitat that may resemble the bare rocks on which the first land plants grew. While the tops of walls are dry in the summer, they are not in the winter, and so this is when mosses' gametes get together. Furthermore, by suggesting that mosses are weak because they need water for their sperm to swim through implies that other plants do not need water, and to suggest that any plants do not need water is a nonsense. Mosses have not found the temporary need for water to be a critical failing; they have been living on land for more than 400 million years.

If these five problems are put together, it is possible to see that one way of solving them simply and simultaneously is to be a liverwort. Gardeners are familiar with these little plants as one species, *Marchantia polymorpha*, that frequently grows on the top of pots in which one has sown seeds. These plants resemble a green liver-like sheet or thallus (though some other species of liverworts have ranks of leaf-like structures along the sides of a simple vein). Like the mosses, the liverworts are tolerant of desiccation. In the upper surface of the thallus, there are barrel-shaped air channels to supply the upper half of the thallus with the carbon dioxide necessary for photosynthesis and the oxygen for respiration. The lower part of the thallus is used for storage of substances such as starch. *Marchantia polymorpha* has unicellular rhizoids that can grow into the tiny fissures in the rocks to hold it in place. These rhizoids do not need to take up water and nutrients because these can be absorbed directly by the plant. As described already, the spores are coated in decay-resistant sporopollenin, and the plants restrict their sexual activity to the wet periods of the year to get round the perceived problem of releasing sperm straight into the environment. One feature of all mosses and liverworts is that they do not grow tall when compared to most ferns, conifers, and flowering plants, and in fact also when compared to many of the green and red algae that preceded them. This is because there is one more problem facing land plants that the seaweeds did not have – gravity. The support provided by the sea to buoyant algae has enabled them to grow into large structures. The air does not provide this support, and the mosses and liverworts therefore stay small, but again it has not prevented them from surviving for hundreds of millions of years. The mosses evolved a very simple internal pipework consisting of hydroids that are sometimes found in fossil plants such as the *Cooksonia* described below.

The question arises, how did plants like *Chara*, or an ancestor of modern-day *Chara*, evolve into *Marchantia* or something like it? We might try to find evidence of an amphibious plant that has a free-floating aquatic form but which can live on mud if the pond in which it lives dries up. Fortunately, there is one such liverwort, *Riccia fluitans*. This could be regarded as a half-way land plant, or missing link, possessing some of the adaptations for living on land in the air but not all in one go. This is, however, pure speculation and not supported by fossil evidence. Sadly, the fossil evidence for the invasion of land and air by plants is generally poor because these soft little plants did not make good fossils, though, as said before, the spores do make good fossils, albeit very small ones.



So the first land plants were liverwort-like and had thrown off the dependence on water as an omnipresent, all-embracing growing medium. This has to be regarded as one of the major landmarks in the evolution of life on Earth because not only did these plants radiate out and evolve into the 400,000 species of land plant that we now see around us, but also they in turn provided many novel habitats for animals, bacteria, and fungi. As ever, evolution did not rest on its laurels but from the liverworts' ancestors evolved many other groups of plants: some extinct, some extant.

## Stems

As plants have evolved, different groups have taken ecological centre stage for differing lengths of time. For many millions of years, the plants were no more than a few centimetres tall. The reason for this was two-fold. Firstly, the plants needed a rigid internal scaffolding to keep them upright against the oppression of gravity. Secondly, they needed pipework through which water and other nutrients could move to ever-increasing heights. In addition, a pump to push the water up the plumbing would help. If you want to gain height, a cell wall and water pressure can get you only so far – a few centimetres tall at best. There is good fossil evidence that 425 million years ago, there were at least seven species of *Cooksonia* (now extinct) around the world which did have stems that varied in diameter from as little as 0.03 to 3 millimetres.

The fossils show that these plants consisted of dichotomously branching stems that terminated with a sporangium. This means that these were the sporophyte part of the life history. (Sadly, there are no fossils of anything that can be considered the gametophyte.) On the surface of these branching stems can be seen stomata for the ingress of gases and the loss of water. The loss of water is not all bad if water is plentiful and if its evaporation can be used to pull water up a narrow capillary-sized tube. In the fossils of *Cooksonia*, there appear to be tubes through which water could be pulled in this way. These are known as xylem tubes, though they may be more like the hydroids found in extant mosses than the xylem in extant vascular plants. It is assumed that the loss of water through the stomata was enough of a pull for water to move up this stem to the top of the plant where it was needed.

However, if plants were to get bigger, they had to come up with something new, or as often happens, have a look in the tool box and see if there was anything already out there in their genetic inheritance that was fit for purpose. What they found was lignin, which was first used by some of the red algae but then not used again. Just what red algae needed lignin for is anyone's guess. This is anthropomorphizing the evolutionary process whereby lignin was included in trees, *but* it is an illustration of one of the major strategies employed by plants, and that is that in order to survive they have to make use of what they have because they do not have the option of running away.

In order to understand the importance of lignin, we need to retrace our steps to the definition of a plant. One of the defining features is the possession of a wall around the cell. This has a number of functions such as restricting the volume of the cell to prevent explosion and preventing the ingress of some molecules. It is not very rigid in its basic form. The primary plant cell wall consists of two layers: an outer layer known as the middle lamella and the primary cell wall itself. The middle lamella is made up of a complex mixture of polysaccharides known as pectin. One of the functions of the middle lamella is to hold cells together in multicellular plants. It is also the part of the plant that breaks its hold when fruit and leaves are shed. Furthermore, it is very important in jam-making. The primary cell wall is made of a mixture of cellulose and hemicellulose. The former is perhaps the most prevalent non-fossil organic carbon molecule on Earth, accounting for one-third of all plant material. Cellulose consists of very long unbranched strands of glucose molecules. The hemicellulose, in comparison, is made up of shorter molecules that branch, and it is made up of more than just glucose. This primary cell wall is strong yet flexible, and when the cell inside it is at full turgor, the whole structure is quite rigid. The flexibility is important because as cells grow the primary cell wall has to grow too. The orientation of the cellulose microfibrils in the cell wall determines the direction in which the cell will grow. However, as we see when a plant wilts, if the turgor of the cell falls due to lack of water, then the plant becomes quite limp and the cell wall bends and folds.

A plant that wilted every time water was in short supply was never going to get taller except in places where it never stopped raining. This was where lignin was brought back in to make up the secondary cell wall. Lignin is strange stuff. It is second only to cellulose in its abundance; perhaps as much as 30% of the dry weight of wood is lignin, and 30% of all non-fossil organic carbon is lignin (until all the forests have been cut down). Chemically, lignin is odd as it lacks a consistent structure. However, this does not matter because it does the job. The job that it is required to perform differs from tissue to tissue and so does its structure. In many parts of the plant, it acts as a cement, filling the gaps between the molecules of cellulose, hemicellulose, and pectin. Lignin makes the cell wall rigid and hard. However, it can also be used as a storage facility such as in seeds, or as waterproofing when it contains a lot of suberin. This has an added advantage and this is that, when compared with the other components of the cell wall, lignin is hydrophobic, making a pipe lined with lignin a much more effective pipe than a pipe lined with cellulose, which is leaky.

Xylem tubes are made up of cells that have lost their living components and are empty. Xylem cells come in two basic designs. The first to appear (and therefore assumed to be the more primitive type) are tracheids. These were followed by the vessel elements. Tracheids are longer and thinner than vessels. They have a lignified cell wall that has a helical reinforcing layer. The tracheids are often associated with parenchyma, sometimes referred to as ground tissue. Where the tracheids join together, they have long wedge-shaped ends where the two flat sides rest against each other. This helps to prevent the continuity of the water stream from breaking as it rises up the stem. Tracheids tend to function in bundles with water leaking sideways from one to another, helping to prevent the inclusion of air bubbles in the stream of water.



The vessel elements are shorter and wider than the tracheids and more expensive to build. They line up very closely and the end walls that connect with the cells above and below them in the stem are perforated, making the passage of the water much easier. There is a variety of perforation patterns, but there is a penalty to pay for the increased speed of moving water and that is that the water stream can break more easily. Vessels were first identified in flowering plants and it was assumed for a long time that this was one of the reasons for the current success of the flowering plants. However, vessels have now been found in a range of plants including ferns, horsetails, and club-mosses, and so, like tracheids, they have evolved several times. Tracheids can hardly be described as a poor relation to the vessels when you consider that the tallest and largest trees have tracheids. The coastal redwoods at 360 feet tall are thought to be right at the limit of current plant engineering performance. It should be noted that the growth habit that we call a tree has evolved many times. A tree is very difficult to define precisely. In law, a tree is a single-stemmed woody plant that is at least 3 inches in diameter, 5 feet above ground level. Botanically, a tree is simply a plant with a stick up the middle – but how that stick is made varies.

Inevitably, with the evolution of these hard and heavy lignified stems came a number of other problems, including not only how do you stand up and also take up water and nutrients fast enough to supply the top of the plant, but just as importantly how do you get the products of photosynthesis down from the top of the plant to the roots where energy is required? Better roots were required, and this is discussed later, but some more plumbing was needed to transport the photosynthates.

When land plants were all mosses and liverworts, there wasn't a great deal of cellular specialization. Most of the cells were photosynthetic and so could look after themselves. Those that were not able to photosynthesize were so close to those that could that simple intercellular transport channels were sufficient. As soon as plants became lignified, the distance between the aerial parts and the subterranean bits was too long for diffusion to work. If the xylem is the ascending transport system, a descending transport system was required, and this is provided by the phloem.

There is one simple difference between phloem and xylem, and this is that phloem cells are still alive. Phloem as a tissue consists of three types of cells. Firstly, there are sieve-tubes, cells through which the sap flows. The cells have no nucleus but they do have a cytoplasm with a few organelles to enable them to fulfil their role. The sieve-tubes line up like sections of pipe. The end walls are like sieves, and through the holes pass larger than average plasmodesmata. Plasmodesmata are small tubes that connect cells to each other, and here they are found connecting the small amount of cytoplasm that there is to the companion cell, the second type of cell in phloem. This companion cell has an above-average number of mitochondria because it is having to power the sieve-cells. There is a specialized type of companion cell known as a transfer cell that has a much-convoluted cell wall, enabling it to gather solutes efficiently from the space in the cell walls around it. The final type of cell found in phloem is the ground tissue in the form of parenchyma, sclereids, and fibres. The sclereids are tough cells that are particularly common in plants in Mediterranean-type environments where the plants experience severe water stress in the summer. The cells in the sclerenchyma are nearly all secondary cell wall with a small amount of cytoplasm at the centre.

Many plants are short-lived, and once they have produced some progeny, they die. Others produce offspring year after year. Some of these perennial plants become woody and live for many years. This requires a stem that not only is resistant to pests and diseases but which is able to increase in girth. This scaffolding does not have to be living, and in many trees it is not. When a tree is felled, it is normally possible to measure the age of the plant by counting the number of concentric rings in the tree trunk. These rings of tissue are produced from a ring of vascular cambium that first develops behind the shoot tip. Cambium is a type of meristematic tissue that develops into other mature types of tissue. This tube of vascular cambium gives rise to xylem in the inside and phloem on the outer side. It also divides sideways to keep pace with the increasing girth of the stem.

In some trees, this does not work because the vascular tissue is not in a continuous ring around the stem, but it is arranged in bundles of xylem and phloem that are scattered randomly through the stem. This is the case in trees such as palms. Here, there is no secondary lateral meristem in the shoot as described in the previous paragraph. There is an apical meristem in the growing tip behind which there are leaf primordia, the meristems from which leaves grow. However, there is an extra structure called the primary thickening meristem. This is small in the shoot emerging from a newly germinated palm seed, but as the stem gets wider, then so does the primary thickening meristem. Its function is to produce vascular bundles, reminiscent of spaghetti coming out of a pasta machine. If the young palm tree is living in optimal conditions, the growing tip will get wider and wider until it reaches the maximum diameter for that species. The plant will then start to grow up and produce a proper trunk. However, if that plant suffers a period of less than optimum conditions in the future, the width of the meristems will decrease, resulting in a narrower stem. However, if things pick up again, the meristems can enlarge back to their former size, with the result that you get a tree with a waist.



When a seed germinates, the first structure to appear is usually the young root because the uptake of water is the first requirement of the seedling. If the seed is thought of as an embryo wrapped in a tough coat with a packed lunch in its pocket, then growing a root first is a good strategy. However, the packed lunch will only support the embryo for a finite period, and there is no going home to mummy for this embryo. It must produce its own food, and to do this it needs to photosynthesize, and to do this it need leaves (or something like leaves).

## Leaves

The arrangement of tissues at the shoot tip is completely different from that found at the root tip. For example, there is no shoot cap because pushing through air should not result in damage. That is not to say that the growing point is not protected in many plants. A simple way of doing this is to have the meristematic tissue surrounded by the young leaves and their stalks. The shoot meristem gives rise to lateral appendages such as leaves as well as more stem, that grows into branches, and terminal organs such as flowers. This is different from roots, as the root does not generally produce any specialized structures just more roots. The control of the development of cells in a root is derived from the position of the cells in the root, whereas in the stem, more sophisticated and complex gene activity controls the operation.

The structure of stems and their cells has already been described, but a stem is basically a support system. Although green stems will contribute to the photosynthetic activity of the plant, it is rarely enough. Only in the case of cacti and some other succulent plants are photosynthetic stems all the plant needs because leaves are too wasteful to be a viable strategy when water is in short supply. However, in the majority of plants, leaves are produced. The arrangement of these leaves is normally constant within a species and is determined by interactions at the growing tip. The shoot meristem is in a constant state of fragmentation and replenishment. As a piece becomes detached from the side, it becomes a leaf primordia that will grow into a leaf with the bud that is found the base of that leaf. The next leaf will be formed when another piece of meristem becomes detached. Where this happens depends on how strong the inhibition is from the previous primordia. This point of minimum inhibition will be at different numbers of degrees around the stem depending on the leaf arrangement of the species in question. In plants with alternate leaves, it will be 120°; if the leaves are in opposite pairs, it will be 180°; and if spirally arranged, it will be approximately 137°.

We probably all think that we know what a leaf is. It is a flat green structure. We might remember from school that it has a thick cuticle on the cells of its upper surface, a thinner cuticle on the cells of the lower surface which is perforated by stomata, and in the middle is an upper layer of cells arranged like bricks (palisade mesophyll), below which there is a more open layer of cells (the spongy mesophyll). Running through all this are veins consisting of xylem, phloem, and probably some lignified stiffening tissue. This is a good generalized leaf structure, but here, as in many aspects of plant biology, there is no true default setting as plants have adapted to so many different ecological niches.

In actual fact, a leaf is a lateral structure on a stem, and in the axil where the leaf comes off the stem there is a bud that may grow into either a vegetative branch or into an inflorescence. The leaf is a determinate structure with a predetermined, finite size. Some 'leaves' are in fact spines (e.g. in some members of the bean family), some leaves develop into pitchers for catching dumb insects (e.g. *Heliamphora*), while other leaves are tendrils for aiding climbing. To be strictly correct, it is the other way around and spines, pitchers, and tendrils may be leaves. This is a simple

**5. The leaves of *Heliamphora* are modified for the capture of insects**

illustration of the principle of homology whereby function does not define a structure, rather identity of a structure is determined by its relative position and developmental origin.

In the same way that there are some leaves that do not look like leaves, there are other structures that look like leaves but are not leaves. The British plant butchers' broom (*Ruscus aculeatus*) has flowers and subsequently fruit growing from the upper surfaces of its leaves. Likewise, a Madeira climber, *Semele*, has flowers growing from the edges of its leaves. When one sees this type of thing going on, one needs to be very careful because all that is flat and green is not a leaf, and flowers cannot grow from leaves. These flowering leaves are simply flattened stems that superficially resemble leaves.

## Branches

The stems of mosses and ferns usually branch dichotomously. This is relatively easy to achieve since it just requires the apical meristematic area to split into two halves. In most seed plants, the shoot meristem stays very much the same size during the growth of the plant, and lateral pieces of meristem become detached to form leaves with the associated buds that in their turn contain a lateral meristem. In some plants, the leaf primordia and lateral meristems leave the shoot meristem together but as distinct structures. In other plants, just the leaf primordia separates from the shoot meristem. It is later as the leaf begins to grow and differentiate that some of the cells revert to being meristematic.

## Roots

It has already been noted that the little liverworts and mosses had little unicellular rhizoids, tiny root-like structures consisting of a single cell, to attach them to the rocks. Rhizoids are found elsewhere in the plant kingdom, including in the Characeae, so again this may be a pre-adaptation to live on land. It has recently been shown that the control of the development of rhizoids in the

**6. *Ruscus aculeatus*, or butchers' broom, appears to have flowers growing on its leaves. These 'leaves' are actually flattened stems**

moss *Physcomitrella patens* is controlled by the same genes that control the development of root hairs on flowering plants.

Tiny rhizoids alone were never going to have sufficient capacity on their own to supply water and nutrients to support a large perennial fern such as bracken, let alone a cactus, redwood, or oak tree. The biology of roots is a much neglected subject; out of sight really is out of mind in this case. Just as the parts of plants above the ground vary from species to species, so do the parts of plants below the ground. One thing that most of them (80% to 95% of all species) have in common is a close relationship with a bacteria or fungus. These arbuscular mycorrhizae benefit the plant by supplying nutrients, in particular phosphates. In return, the fungus takes sugars from some of the cells in the root.

This type of relationship appears in the fossil record and seems to have evolved on the underground parts of plants that we do not consider to have true roots. It is now thought that roots have evolved at least twice, not including rhizoids. The first roots are found in fossils 350 million years old. The plants to which these roots belonged are in the group known as the lycopsids. This is a group on a branch of the evolutionary tree between mosses and ferns. The lycopsids were the first plants to grow into proper trees and so must have had a root system capable of supporting the crown of a tree. The fossil roots of these ancient lycopsids differ from more modern roots in that there is a single central core of xylem and phloem surrounded by parenchyma, and an epidermis with root hairs.

By the time the ferns were producing roots, a different developmental pathway had evolved. In the majority of ferns and seed plants, the root tip consists of four different regions. A root cap at the tip protects the more delicate structures from damage as the root pushes through the soil. Behind this is the meristem where the cell division happens. In the centre of this meristem is the quiescent centre which is believed to control the timing of the differentiation (as opposed to division) of the dividing cells that surround it. The new cells produced by the meristem then become the elongation zone as the tip grows away from them. Having expanded in the elongation zone, the cells then find themselves in the differentiation zone, where they develop into epidermis, cortex, endodermis, or vascular tissue. Further back, a tube of lateral meristem may be produced in perennial plants. This cambium will give rise to secondary xylem (on the inside of the tube) and phloem (on the outside), thus resulting in bigger permanent roots. These roots then will branch and form the complex root systems seen when high winds uproot mature trees. While this type of root system is common, there is another type known as adventitious roots. These grow from a part of the plant *other* than the primary root that grows from the embryo in the seed. This is the situation in the monocots and accounts for why mature palm trees can be transplanted with very little root. This will be explained in more detail in Chapter 5.

So now we have a diverse range of land plants ranging in height from ground-hugging liverworts to lofty coastal redwoods, and from tiny floating duckweeds to colossal eucalyptus trees. However, no matter how big or small, they share one goal in life: to make more plants for after they are dead. This they do on many variations of the previously described alternation of generations. Time for Chapter 3.

# Chapter 3
# Making more plants

The production of more new individuals resembling their parents is one of the defining features of living organisms and plants are no exception to this rule. When plants first colonized land and air, they had a reproductive system in place that worked in water, but could it still function where water was not ubiquitous? Obviously, the answer to this is yes, but some things had to change.

There are two fundamentally different ways to make more plants that we see in most, if not all, of the major groups of land plants. The more simple way is to make a replica vegetatively. The advantages of this method are that it is simpler than sexual reproduction, only one plant is required to make more, it is quicker, and if the parent is well adapted to its environment, then so will the offspring be. We find vegetative (or asexual) reproduction as a common trait of invasive species, so it is clear that there are powerful, short-term advantages to this strategy. There are a number of ways in which plants can produce free-living replicas of themselves.

## Asexual/vegetative reproduction

Liverworts have a structure known as a gemmae. This is an undifferentiated blob of cells that grows in the gemmae cups on the surface of the thallus. When the blob is ready to be released, it is dislodged by raindrops and is washed away. When it comes to rest, it grows into a plant like its parent. Mosses like *Aulacomnium androgynum* produce gemmae from the tips of their shoots. These gemmae are not amorphous blobs but have distinct ends with meristematic apical and basal cells with several cells in between. The dispersal of these gemmae is by any means available.

Anyone who has been fell walking in the English Lake District should not fail to be impressed by the ability of bracken (*Pteridium aquilinum*) to spread. As the plant grows across the hillside, the individual shoots are connected to the parent plant. However, these shoots each have their own root system and are quite capable of living on their own. A few other ferns produce tiny plants from their leaves. *Asplenium viviparum* and *Asplenium bulbiferum* are two such species.



Vegetative reproduction is relatively rare in gymnosperms, but a well-known example is found in the coastal redwood (*Sequoia sempervirens*). This species produces young plants in a ring around its base. When the parent plant dies, it leaves a concentric ring of young trees. Eventually, these will each produce another ring of replacements. It could be argued that this is regeneration rather than propagation of the plants, as there is no hint of dispersal. The Huon pine (*Lagarostrobus franklinii*) from Tasmania also produces these long-lived clonal populations which have been claimed to be more than 10,000 years old and so are the oldest organisms, or genotypes, on Earth.

As all gardeners will know, vegetative reproduction is very common in flowering plants. Couch grass (*Elytrigia repens*), ground elder (*Aegopodium podograria*), and bindweed (*Convolvulus arvensis*) are just three plants that are very successful weeds of cultivated places because of their ability to regrow into mature plants from just a short fragment of plant that is left in the soil following digging. As mentioned before, many of the world's most invasive plant species take over an area by rapid vegetative spread. Plants in this category include *Rhododendron ponticum* and brambles. In the former, the lowest branches will root as they touch the soil, and in the latter the shoot tip will develop roots if it is in contact with soil for a few weeks. A particularly insidious version of this form of reproduction is found in false garlic (*Nothoscordum inodorum*) which produces tiny bulbs, or bulbils, from around its main bulb and also near the flowers in the inflorescence. This, combined with prodigious seed production, makes this an impossible plant to eradicate from where it is not wanted. The water lettuce (*Pistia stratiotes*) produces perfect little plants on the ends of lateral shoots. These plants are broken off by disturbance by large animals. However, if left undisturbed, this species will form huge mats that can become floating islands strong enough to support a rhinoceros (albeit a small rhinoceros).

**7. *Nothoscordum* produces many small bulbils from around the parent bulb, making it a very persistent weed**

In some plants, there is a strange half-way-house between sexual and asexual reproduction known as apomixis (or sometimes parthenogenesis, a term more commonly used when animals are being discussed). In apomixis, seeds containing a viable embryo are produced without the egg being fertilized. This means that the seeds are genetically identical to the mother plant. The problem for botanical recorders is that each region of the world may have slightly different clones of the species. This is the case in dandelions, for which there are more than 250 named micro-species that are simply clones. These clones may be short-lived, but there are implications here for conservation. Do we try to conserve each variant, or should we try to preserve the process by which this number of micro-species is maintained?

However, being well suited to a particular habitat today is no guarantee of future success if the habitat changes. While evolution is incapable of foresight, it can be seen that the ability to produce variation, through mutations and reassortment of genes during sexual reproduction, has provided the modifications that Darwin recognized as being central to evolution. As Darwin suggested, 'it is a general law of nature that no organic being self-fertilizes itself for an eternity of generations', because this leads to something known as 'inbreeding depression' whereby progeny become weaker and weaker. This being the case, we come to the second, more complicated, means of reproduction, and that is the fusion of a sperm and an egg in sexual reproduction.

## Sexual reproduction

In Chapter 1, the life history of plants was introduced, and it was shown that this involves a strategy unique to plants – namely, the alternation of a diploid and haploid stage known as the alternation of generations. If we accept that the green algae like *Chara* are the closest relatives of the land plants, and if we accept that liverworts are the living descendants of the lowest branch on the evolutionary tree for land plants, then the differences between these two groups of plants are a clue as to what changed to enable plants to colonize land and air. In Chapter 2, we have seen how many of the problems were solved (desiccation, support, finding raw materials and nutrients), but there is one outstanding problem, and that is how to bring gametes together when the previous means of swimming through the omnipresent water is not always an option.

To recap very briefly, the *Chara* plant produces lots of sperm with two tails, or flagella, that swim to another plant where lots of eggs are sitting in a structure known as an oogonium. The sperm fertilize the eggs (one sperm per egg) and a zygote results. The zygote then undergoes meiosis to produce four spores. Sperm, eggs, and spores contain one set of chromosomes (haploid), whereas the zygote has two sets (diploid). There is one very big difference between this life history and all of the land plants, and several smaller differences. The very big difference is that the zygote of land plants does not immediately divide meiotically. Instead, it divides mitotically (simple cell division) and grows into an embryo. This is new, and this is why the land plants are referred to as the embryophytes (by botanists mainly!).



## Mosses and liverworts

So we start with the liverworts, and in particular with *Marchantia polymorpha* as this is probably the most studied and best understood liverwort. The plant of *Marchantia* is a dark-green thallus, shining like wet liver and bifurcating as it spreads across the surface of the moist growing medium. On its surface, there may be the gemmae cups, each about 5 to 8 millimetres in diameter. However, at certain times of the year, wet times, umbrella-like structures growing from the surface of the thallus can be seen. There will be some umbrellas that have an intact top and some that will look skeletal as if all the material has blown away and only the spokes remain.

From the upper surface of the intact umbrellas, sperm will swim from the antheridia. These antheridia (singular, antheridium) are one of the smaller differences between *Chara* and the land plants, though they don't *look* very different from the sperm-producing organs of *Chara*. The antheridia are one of the two types of gametangium found in land plants. (The suffix -angium, plural, -angia, is used to describe a reproductive structure, so gametangia produce gametes and sporangia produce spores.) The other type of gametangium is the one that produces the egg – the archegonium; the archegonium is the other small difference between *Chara* and the land plants, though they don't *look* very different from the egg-producing organs of *Chara*. A significant difference is that the archegonium produces just one egg, while each oogonium of *Chara* produces lots of eggs.

So the sperm swims using its two flagella towards an egg in an archegonium. These archegonia are on the underside of the umbrellas with just spokes. The plants will be either male or female and so just one type of umbrella. The mechanism that determines the sex of these plants is genetic and is due to heteromorphic chromosomes: a small Y (male) chromosome and a large X (female) chromosome. This might sound familiar, and indeed in humans males are XY and females XX, but you must remember that the liverwort plants producing gametes (the gametophytes) are haploid, so the males are just Y and the females just X. Meanwhile, back at the liverwort archegonium, the egg has been fertilized by the sperm and a diploid zygote formed. This zygote now divides and grows into an embryo. It is *not* released by the archegonium but derives its nutrients from the female gametophyte. The embryo grows and develops a sac hanging on a stalk; this is the sporophyte of the liverwort. It is not free-living but dependent upon the gametophyte for all its 'life'. Inside the sac, spores are being produced and wrapped in sporopollenin. This is the tough outer layer described in Chapter 1. The sac is therefore the sporangium. The spores are produced following meiotic divisions that reduce the number of chromosomes from diploid back to haploid. The spores all *look* the same, and so this plant is said to be homosporous. In reality, the spores are not all the same because half are female and half are male, but they are the same size and are produced by the same sporangium.

These spores fall from the ruptured sporangium and are blown away by the wind or washed away by water. The spores are tough little structures because of their coat of sporopollenin that they inherited from their aquatic ancestors. When the spores come to rest on a suitable medium and the temperature and humidity are right, they germinate. They grow into either a male thallus or a female thallus, and these gametophytes then grow until mature, when they will put up their umbrellas and the life history is complete.

The life history of the mosses is similar but different. We already know that mosses are one step ahead of the liverworts because they have stomata. We may be more familiar with mosses as they are common plants, in towns growing on walls and in gardens growing in lawns. The plants growing on walls are more useful to consider here because at certain times of year, particularly the late winter/early spring, the plants sport little structures that look like periscopes – a thin filamentous stem supporting a capsule a few millimetres long. This capsule is the sporangium and so out of its open end come tough haploid spores, and again they all look the same. We are ignorant about the determination the sex of the gametophytes of mosses, but in *Ceratodon purpureus* it has been shown to be genetically determined probably by sex chromosomes. However, in some mosses like *Physcomitrella patens*, the spores germinate and grow into a bisexual (or hermaphrodite) gametophyte.

Irrespective of whether the gametophyte is bisexual or single sex, antheridia and archegonia develop, often near the tips of the leafy shoots. When the weather is wet, the sperm swims from its antheridium to an archegonium, where it hopes to find the egg of its dreams. How it finds the archegonium is uncertain, but in some species there appears to be a chemical signal being emitted by the archegonium. Once the sperm has fertilized the egg, the zygote grows into a multicellular structure that grows into the sporophyte. In the case of the *Tortula muralis*, that sporophyte is the periscope that we see on the tops of walls in late winter. Within the capsule, meiosis takes place, haploid spores are produced, and the life cycle is completed.

## Ferns

The next major group of extant plant species on the tree of life are the ferns. If you look at the underside of the older leaves of ferns, you will often see erumpent pustules releasing a brown dust. If you cut the leaf off the plant, secure it gently on to a piece of white paper, and put the paper on top of a boiler or Aga overnight, when you remove the fern leaf the following morning, you will have a perfect representation of the distribution of the sporangia on the fern leaf, because that brown dust is thousands of identical spores all coated in sporopollenin. This means that the erumpent pustules are the sporangia and the leaf is part of the sporophyte. This is obviously rather bigger that the periscope of our little moss growing on the top of the wall, and it is different in another way: this mature fern sporophyte is free-living, whereas the periscope moss sporophytes were completely dependant on the gametophyte underneath them.

So the spores are released and are blown away, hoping to land on a piece of wet soil or a wet branch, or anywhere in fact with reliably high humidity and 'soil' with some nutrients. The spores germinate and grow into the gametophytes. The gametophytes are free-living and green, but they are not large. They are rarely more than 10 millimetres in any direction. They are often flat and thallose, not dissimilar in many cases to a small liverwort, and they can be mistaken for liverworts since they often grow together in the same conditions. The gametophyte may be male, female, or bisexual depending on the species or on the conditions. In *Ceratopteris richardii*, the first spores that germinate are bisexual, producing both archegonia and antheridia. However, these pioneering gametophytes produce a hormone that affects the development of any newly germinated spores nearby, and if the hormone reaches these young gametophytes in first few days (two to four days), the gametophyte is totally male and produces just antheridia and thus just sperm. There is an advantage to this strategy in that it promotes outbreeding. Another way to promote outbreeding is for the gametophyte to produce the gametangia at different times, thus being male for a while and then female, or vice versa.

A problem inherent in the gametophyte producing both antheridia and archegonia is that it could produce just self-fertilized eggs and thus zygotes that are a diploid clone of the gametophyte. Among Charles Darwin's portfolio of biological one-liners is 'Nature abhors repeated inbreeding', and, as we shall see later in this chapter, Nature goes to a lot of trouble to avoid inbreeding – most of the time. 'Most of the time' because one of the challenges facing all plants, be they aquatic or terrestrial, is that they are immobile. If they are the only member of their species in the vicinity, they are not going to be able to produce any young at all, if they have to be fertilized by sperm from a different plant. So many plants have a policy of accepting their own sperm if there is no alternative, so that at least they get another generation and another chance to outbreed. Darwin called this strategy 'reproductive assurance', and it is widespread.

Leaving aside concerns about where the sperm have swum from (and they now have many flagella to help them on their way), a zygote is formed in the archegonium, and this grows into an embryo. However, rather than growing into a sporangium on a stalk of some type, the embryo produces a little leaf above the gametophyte and a little root below it. This young plant is initially supported by the gametophyte, but by the time it has a few leaves and a few roots, the parent gametophyte has been drained of all its resources and is an empty vessel – a possible metaphor for the human condition.

So far, all of the sporophytes in this story have produced just one type of spore; they have been homosporous. It is believed that fossil plants like *Cooksonia* were also homosporous. However, many plant groups and the vast majority of plant species currently extant are heterosporous. Heterospory has evolved several times, and the first time may have been before the ferns came to prominence. Between the mosses and the ferns on the tree of life are the lycopsids. This rather enigmatic and diverse group of plants was among the first to grow into trees and also the first to produce male and female spores. Under no circumstances must these spores be confused for gametes. They are spores, so like gametes they are haploid, but all a spore can do is grow into a gametophyte. Spores cannot fuse with each other to give a diploid zygote.



## Seed plants

The last major group of land plants are all heterosporous, and these are the seed plants: the gymnosperms and the angiosperms. The gymnosperms include the conifers, the cycads, Ginkgo, and a small bunch of botanical weirdos including *Welwitschia mirabilis*, the iconic oddity from the Namib Desert, and *Ephedra distachya*, the original source of medically exploited ephedrine (described in more detail in Chapter 6). The angiosperms, on the other hand, are the flowering plants. It is believed that the seed plants have a common ancestry, and it is assumed that their most recent common ancestor was heterosporous, but what it looked like is the subject of several careers in scientific research and inspired guesswork. The emergence of seed plants may be a mystery, but the origin of the angiosperms is an 'abominable mystery' according to Darwin, and it is not yet solved.

**8. *Lycopodium* in Japan. Relatives of this plant were among the first plants to grow into trees**

Evolution is a complicated business. It has no foresight, so it simply retains what works today and does not have an attic full of things that might be useful in the future. It is also quite uncharitable. Darwin believed that no organism would acquire a trait for the benefit of any organism but itself. It may be that what is advantageous to one organism might be exploited by another at some point in the future, but that will not be considered by evolution. The seed is a good example of this. It is difficult to think of the world being as it is now had not seeds evolved on seed plants. Many animals, including 6.5 billion humans, depend on seeds for their staple food. Yet seeds can only have evolved if they gave the plants that possessed them an advantage. Those advantages might have been survival and dispersal, and perhaps the ability to survive dispersal. So seeds permitted transport in both space and time. But what is a seed, and how is it produced? Seeds have been described already in Chapter 2 as 'an embryo wrapped in a tough coat with a packed lunch in its pocket'. If the seed has an embryo in it already, then it is formed *after* reproduction has happened.



If you have ever parked your car under a cedar tree (*Cedrus* spp.) in the spring, you may well have returned to find it covered in yellow dust. You might even have found some soft, sausage-shaped cones on the ground around the car. If you look into the tree, you will see more of these small cones and some very big woody cones a few inches in diameter. Inside these woody cones is where this part of the story begins because here we find a sporangium. (It is in fact known as a mega-sporangium because it produces larger spores than the other type of sporangium found on a seed plant.) These spores are not released as has happened in every species we have looked at so far. The megaspore is retained in the cone, where it grows into a female gametophyte that is similar in size to the fern gametophytes. An archegonium grows on the gametophyte, and so again this is not so different from the ferns, except that this gametophyte is not free-living but is totally dependent on the sporophyte for its nutrition.

Now if this has made any sense, you should be asking yourself where is the male gametophyte? Well, the male gametophyte is made somewhere else and then has to find the female. The male gametophytes grow from microspores produced by the microsporangium and these sporangia are found in the other cones, the cones you found on your car bonnet. The microspores are produced with the traditional coat of sporopollenin. However, before they are released, the spore inside undergoes a bit of cell division, and in addition two large air sacs grow on the outside of the sporopollenin. This whole structure is an immature male gametophyte. It is immature because it has no means of dispensing its sperm yet. This immature male is hoping to find a mature female gametophyte. The problem is that rather than the two gametophytes sitting next to each other on the soil, as was the case with the ferns, mosses, and liverworts, the females are sitting in a woody cone elsewhere on the tree, or preferably on another tree.

The ingenious males find their females by gliding, buoyed up by their air sacs. The hope is that one of them will land on the woody cone, preferably on another tree. There is a million to one chance of this happening, but this is enough. These gliding immature male gametophytes are much better known as pollen grains. When it lands on a woody cone, the pollen grain takes in water from the imaginatively named pollen drop. It then grows a tube and at the tip of that tube is a sperm cell. This sperm is not produced from an antheridium; these are now a thing of the evolutionary past. This sperm has no flagella, so the tube has to grow to the neck of the archegonium on the female gametophyte that has been waiting patiently for its male caller. The sperm fertilizes the egg, and a zygote is produced that grows into an embryo; so far, so good, except that the embryo is nowhere near a suitable place to grow, so the development is halted, a thick coat is provided by a part of the woody cone, and the female gametophyte takes on the role of 'food for the journey'. The seed is released from the woody cone and floats to the ground, hopefully some distance from its mother. Once there, the embryo resumes its growth and grows into a cedar tree.

If this cedar tree had instead been an oak tree, or a magnolia, or a daffodil, the process of bringing the sperm and eggs together would have been different again. These flowering plants do produce pollen, but rather than the microsporangia being situated in a cone, they are in the anthers on the end of the stamens in the flowers. The pollen is similar to that of the cedar tree (and other gymnosperms) in that it is an immature male gametophyte on the look-out for a mature female gametophyte. The problem for the flowering plant's pollen is that the female gametophyte with its egg has been hidden inside the carpel. This carpel is the major innovation in the evolutionary transition from the gymnosperms to the flowering plants. The carpel consists of three parts: the stigma (the landing-pad for the pollen); the style (a stalk, of variable length depending on the species, that connects the style and ovary); and the ovary that contains the female gametophyte(s), each with its own egg. The female gametophytes have developed from the megaspores. These megaspores have been produced by the megasporangia in the ovary. It has to be noted that these female gametophytes are a sad apology for a gametophyte compared to everything that has been produced by other land plants. These angiosperm female gametophytes consists of just a few haploid cells (one of which is the egg cell that has been produced without the aid of an archegonium) and one cell that has two haploid nuclei. It is easy to make a case for this binucleate cell being awarded the prize for being the most important cell in the world, because it develops into not only food for the embryo, but it is the food for the world in grains of rice, wheat, and maize.

So the pollen lands on the stigma. How it does that with precision is one of the best stories in plant biology and will be dealt with at the end of this chapter. For now, we can assume that the immature male gametophyte has landed on the stigma, and its problems are just beginning because this is 'meet the mother-in-law'. The stigma has two questions for this young male. Firstly, are you the right species? Some mothers are less fussy than others, or just visually challenged, and this is why sometimes pollen from another closely related species gets through and hybrids may be produced. This can be a very important way of producing novelty for evolution to exploit or reject. The second question is, are you me? As with all organisms, inbreeding is second-best for plants. Again, some plants are much more diligent in this respect than others. Plants in the cabbage family, for example, are very keen to avoid being fertilized by their own sperm. Once the pollen has answered the questions adequately, water is passed to it and the tube grows down the style. More than one pollen grain may have germinated, and there is evidence that there is now a race to the eggs and that selection for particular traits may happen at this stage.

When the pollen tube emerges from the end of the style into the ovary, it grows towards an egg and releases not one but two sperm. One of these does the decent thing and finds an egg and makes an honest zygote out of it. This zygote grows into an embryo that then has its development arrested, as was the case in the gymnosperms. The second sperm, however, is not just a spare. This has to find the cell of the female gametophyte that has two haploid nuclei. The sperm fuses with this oddity, with the result that the cell is triploid, that is, it has three copies of each chromosome. The reason why this might be considered 'the most important cell in the world' is because this goes on to grow into the endosperm, which is not only the food supply for the embryo but also the food supply for *Homo sapiens*. When we eat anything made from rice, wheat, or sweetcorn, we are eating triploid endosperm.

So how does the pollen of flowering plants get moved with precision? Well, in some cases, it does not and wind is employed with the usual long odds. Rarely, water is used, but more often an animal is hired if a suitable reward can be provided. The rewards are varied. Provender (nectar, starch, and pollen itself), a bed for the night, a bed for the night with someone else, somewhere to lay your eggs and to raise your young are the rewards on offer. The size of the rewards vary depending on the size of the visitor, so a bird or bat requires a bigger reward than a midge. Attracting the pollinator is important, but animals are straightforward and so attracting them is simple. Smell, shape, and colour are all used. Some plants have become embroiled in very close relationships with just one pollinator, which is fine so long as nothing happens to the pollinator. Some species do not care who comes calling just so long as someone does. This is going to increase the risk that the pollen lands on an inappropriate stigma, but at least the plant has been visited.

It is often possible to match pollinator to flower and flower to pollinator, and the study of pollination syndromes was the part of Charles Darwin's experimental work that gave him the greatest pleasure. He wrote a great deal about the various contrivances that plants employ. However, it has recently been shown that this, like so much of plant biology, is not as precise a relationship as one might assume. Some seemingly generalist flowers with no specific adaptations are still only visited by one pollinator, and conversely some seemingly well-adapted flowers are visited by many different organisms. Likewise, although it is possible to describe your perfect bee-pollinated flower or bird-pollinated flower, these are rarely found in nature when you look for them.

So pollination, or the distribution of immature male gametophytes, is a critically important stage in the life of seed plants. It helps to stir up the gene pool as well as to distribute genes. In the seedless plants, the flagellated sperm has to take on this responsibility. As this has worked for more than 450 million years, it can hardly be seen as a problem. However, there is another problem for which neither a flagellated sperm nor a pollinator is suited, and that is the dispersal of the new embryo away from its parents. The seed plants have seeds but the seedless plants have another way, and this will be looked at in the next chapter.

# Chapter 4
# Moving around

If you live in the water, there are two ways to get around: drift or swim. Some of the glaucophytes are swimmers, some of the red algae are drifters (such as *Polysiphonia*), and some of the green algae are swimmers (such as *Chlamydomonas*, *Volvox*, and *Ulva*) and some are drifters (such as *Spirogyra* and *Chara*). The swimming referred to here is that of spores (*Ulva*) or the whole organism (*Chlamydomonas* and *Volvox*). The swimming of sperm does not count, as this is not vegetative dispersal, it is genetic dispersal of the male germ line but nothing more. A sperm cannot develop into a free living organism. The swimming spore of *Ulva*, on the other hand, is able to grow into a free living plant. Is dispersal of plants that live in water important when compared with land plants? Is dispersal important for land plants? The answers to these questions seem to be coloured in some societies by the human condition. It is assumed that children will leave the family home in some regions, while elsewhere they are expected to join the family team and support the group. Plants are asentimental, and dispersal mechanisms will be selected for only if they increase the chance of the plant's genes getting into the next generation.

So back to the questions. Is dispersal of plants important? The aquatic environment is more homogenous than the land. The abrupt barriers that exist on land are rarer in water; gradients exist but they tend to be gradual. Swimming can be useful in this type of environment because it enables young plants to get away from other members of their species with whom they will be competing more strongly than any other organism. However, and this applies to land plants just as much, if your parents are growing well in one place, why risk leaving the area? Perhaps this is why as many as 80% of land plant species simply drop their seeds at their roots and make no attempt to promote their dispersal. Compared to the aquatic environment, land is spatially heterogenous, and so we might expect to find more land plants investing in dispersal mechanisms. However, in addition to spatial heterogeneity, land can be very temporally heterogenous, so we need to consider the dispersal of plants in time as well as space because plants can enter suspended animation in a way that animals can only dream about.



Let us first consider the liverworts and mosses. These plants produce haploid spores that grow into free living gametophytes. We see these plants not only at ground level but also on trees, on high roofs and walls and occasionally in the window ducts of old Land Rovers. The spores are light enough to be blown to any height and, thanks to their coat of sporopollenin, to survive the journey. The same is true to a certain extent for ferns. There are many ferns that do not live in soil. There are epiphytic ferns that grow on the outside of branches, some are free-floating and some grow between the stones in walls. For the epiphytic species, the spores must be able to reach the branches of the trees upon which they are to grow. This is equally true for epiphytic flowering plants such as orchids, many bromeliads (the pineapple family), and some members of the arum family, but more of these later. So from the most casual observations, liverworts, mosses, and ferns seem to have no trouble surviving and spreading. Furthermore, the spores are so small that they are not going to be a very attractive food source. This cannot be said for seeds, so what is so special about seeds? Why would evolution favour the production of such an energy-expensive structure as a seed that would become the favourite food of many organisms?

There are very few, if any, rules in seed biology. There is still a great deal of research to be carried out before we can say that we understand the ecology of seeds. One problem is that there are at least 350,000 species of seed plant and they are all different. Species within a genus can differ in their seed biology, and even plants in the same species can vary in different localities. For example, plants of *Pinus brutia* grow from Crete to northern Greece. At the south of its range, it never experiences frost, whereas up north nights frequently fall below zero in the winter. The seeds of the southerners show no dormancy and will germinate when the autumn rains begin, like many plants in Mediterranean-type regions of the world. On the other hand, the seeds from northern plants do not want to germinate just before a harsh winter, and so their seeds require chilling before they will come into growth in the spring. These are still the same species because acclimatization is not the same as speciation.



One thing we can state without fear of contradiction is that seeds are much smaller than their parents and that they are able to withstand harsh conditions that would kill their parents. It is this ability, perhaps more than any other, that influences the various strategies seen in plants that produce seeds. For those plant species that flower once in their life and so produce just one batch of seeds (monocarpic species), it is critically important that at least one of their seeds survives to flowering size to take their genes to the next generation. It will probably be stated many times in this chapter, but it bears repetition, that the vast majority of seeds do not result in a plant that produces more seeds. If you are an oak tree and have many seasons of seed production, this is not an issue, but for an annual you only have one chance.

Each of the 350,000 species of seed plants is different from the others. Each individual within a species has choices to make during its lifetime, leading to even more diversity. Biologists sometimes try to reduce biology to the allocation of resources in order to rationalize the behaviours and actions of living organisms. So when plants produce seeds, they have to balance the need to grow with the need to put energy into the production of those seeds. It is clear to see that it takes a lot out of a plant producing seeds because in those species where the male and female flowers are on separate plants, the females tend to be smaller and shorter lived than the males.

Do you produce many small seeds or fewer bigger seeds? If you produce lots of seeds, does this increase your chance of survival? The fact that monocarpic species tend to produce very large numbers of seeds implies that the probability of survival increases the more seeds that you produce. Do you use fats or carbohydrates as the energy supply in the seeds? Fats are more expensive to make but are richer in energy gramme for gramme. Plants put between 2.3% and 64.5% of their resources into seed production. There are very few rules in seed biology.



Another issue is how long to take to reach maturity. Unlike animals, sexually active plants of the same species can and do vary greatly in size. If highly disturbed, plants tend to rattle through their life history fast before the next cataclysmic change. In more stable places, seed production may be constant. For example, fig species, which all have a unique relationship with a different species of fig-wasp, produce flowers all year round to satiate the wasp. Beech trees, on the other hand, demonstrate a different strategy, that of masting, whereby a very large number of seeds is produced approximately every seven to ten years. It is believed that the advantage of this is to produce so many seeds that there are far more that the natural predators can consume. Oak trees, on the other hand, produce many acorns each year, many of which get buried by squirrels and then forgotten about. This appears to be good news for the oaks.

9. ***Lodoicea maldivica*** **has the largest seeds in the world. It grows only on the Island of Praslin**

Seeds vary greatly in size. The seed of orchids is tiny, almost dust-like, whereas the seeds of the double coconut, *Lodoicea maldavica*, is the size of two Aussie Rules footballs; that is, a variation of seven orders of magnitude in weight, and yet both the orchid seed and the coconut seed fulfil the same function, and both have the potential to produce a mature flowering plant. Even on the same plant, seeds can vary in size by three orders of magnitude, and the seeds within the same fruit can be different. One rule of thumb seems to be that plants that live in shade tend to have seeds from the bigger end of the scale, perhaps as this enables them to become established in an environment with lower energy levels. Seed size appears to be linked to neither the moisture content of the soil nor the nutrient status of the soil. It may just be that by producing seeds of various sizes, the plant is hedging its bets and increasing its chances. One size of seed does not fit all situations.



In the previous chapter, we saw how many plants have come to rely on a stranger to distribute their sperm, and those who do not trust an animal have thrown caution literally to the wind. We know that neither biotic nor abiotic pollination methods are totally satisfactory because hand pollination of a plant is nearly always more successful than leaving it to nature. Some families of flowering plant are particularly hopeless in this respect. Only up to 7.2% of flowers in the Protea family are successfully pollinated, though far more are visited by the pollinator. It may be that their chosen pollinator is now extinct. (Plant species have a general problem when forming joint ventures with animal species, and this problem is that plant species last 30 times longer in evolutionary terms than animal species. So if you are going to use an animal to distribute your pollen or your seed, you will have to be prepared to change suppliers repeatedly.)

We see in plant reproduction a good example of the Allee effect, namely that it is better for the members of a species to hang around with each other – the scientific safety in numbers principle – because the fewer plants and flowers there are, the fewer successful pollinations there are, and so the fewer seeds are produced. Before you start to think there is a rule appearing, you have to remember that although the bigger groups of inflorescences attract more pollinators and so produce more seeds, they also attract more flower grazers, more seed predators, and more fruit grazers. So there is no optimum size for seeds; the best size is time and space specific, and so the best strategy is variation.

Charles Darwin was not only interested in pollination, he was also fascinated by the geographical distribution of species. He carried out some elegant experiments at Down House to discover if seeds of land plants could survive the salinity of sea water for long enough to enable them to float from one continent to another. He suggested the feet of birds as another means of long-distance dispersal, particularly to oceanic islands, and this is one of the first things that children learn about in school nature lessons. They are shown how the fruit of goosegrass (*Gallium* spp.) and teasels (*Dipsacus fullonum*) adhere to the fur of animals, and because it is the first thing that we are shown it is assumed that this is how most plants disperse their seeds. The truth is that fewer than 5% of species travel on the outside of animals. These species are low-growing (that is, animal height) and from many different habitats. Wind dispersal is also assumed to be common after we have dropped a sycamore 'helicopter', and yet very few structures improve the lateral movement of seeds, they just make them fall more slowly, allowing the horizontal wind to have more effect. Wind can lead to extraordinary dispersal if it is very strong, but it has been shown that generally a forgotten cache of seeds will be further from the parent plant than a wind-dispersed seed. So animal fur and wind are responsible for aiding seed dispersal but in a small minority of species. Some parent plants physically eject their seeds using a ballistic mechanism. The most spectacular of these may the squirting cucumber, *Ecballium elaterium*, that releases its seeds as a stream of exhaust as the fruit breaks off from its stalk. This is so efficient that every seed is ejected during the short flight of the mini-cucumber.

The main alternative to these two is the seed being eaten by an animal and being excreted by that animal some distance from the mother plant. This is a risky strategy as the animals (normally a mammal or bird) have eaten the seeds for their nutritional value and they will grind up and kill the seeds in order to extract that food. Interestingly, those seeds that are good at surviving a trip through the gut of sheep are also those species that persist for many years in the soil seed bank. One way to reduce this risk of being digested is to coat the seed in a laxative; this is more common in bird-distributed species. This may be a good strategy for the plant, but it does make the study of bird dispersal difficult because the dispersal is random and difficult to map! In the same way that pollinators and flowers can be paired up, so can seeds and animals. Birds are attracted to scentless, brightly coloured seeds – reds, blues, and blacks are often combined – whereas mammals go for something smelly and tasty but dull-coloured. The actual evidence to support the dispersal of viable seeds by animals is thin. In two studies, 40,000 deer pellets and 1,000 kilograms of rhino dung were inspected and no viable seeds were found. In another study, of 40,025 seeds eaten by house finches, only 7 survived the experience.

Evidence for the impracticality and foolishness of using the gut of an animal as a vehicle can be found in the fact that many seeds and fruits are toxic. As with many aspects of plant biology, there are many reasons why a seed or fruit might benefit from being toxic and they are by no means mutually exclusive. A tasty but laxative fruit will ensure swift and safe passage of the seed through the gut. An emetic may ensure swift and safe passage of the seed out again before it reaches the gut. If toxic, it might kill the animal, giving the seedling the nutrients of the decaying host. The toxins might be there to inhibit the germination of the seed – physiological dormancy, of which more later. The toxin may be

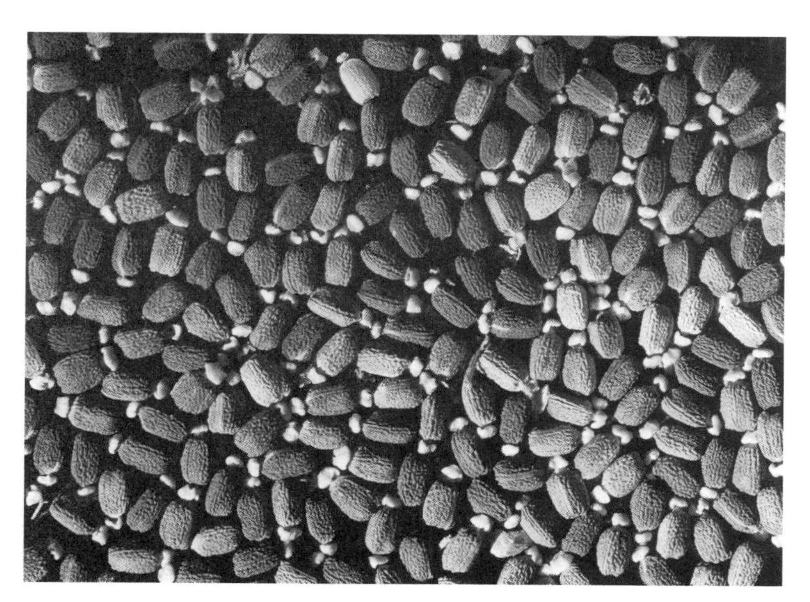

**10. Seeds of *Euphorbia stygiana*, showing the elaiosome**



toxic to seed predators but not to fruit dispersers. A good example of this is the English yew, in which the red aril is safe to eat but the seed is toxic. Finally, the toxin may be a defence again pathogens.

One group of animals that are important in seed dispersal, particularly in the Mediterranean-type regions of the world, are ants. Ant-dispersed seeds possess an elaiosome. This body of fat, usually much smaller than the seed, is an intoxicant of the ant in so far as it stimulates the ant to pick up the seed and take it back to the nest, where it bites off the elaiosome before taking the food inside the nest, leaving the intact seed outside. The soil around the nest is often higher in nutrients than nearby, and often the seeds are taken a couple of inches below ground, where they are protected from not only other predators but from the harmful effects of fire that is common in these areas. Many species of plant are dependent on the ants for this protection, and so when non-native ant species drive out the native species, the plants are left exposed and vulnerable to local extinction.

The problem of non-native species will be dealt with in more detail in Chapter 7, but it is relevant now because the main reason why plants grow here and not there is that they cannot get there. Humans, however, have broken down every dispersal barrier, accidentally and on purpose. Farmers have moved seeds around in contaminated seed sacks, in manure, and on stock – 400 sheep have been shown to move 8 million seeds per annum. Gardeners have moved thousands of species well beyond their natural ranges. Finally, foresters have introduced more productive trees into many countries. The damage that these non-native species can do must not be under-estimated.

Seed dispersal is 'good' for several reasons. Firstly, the parent plant will have become a mecca for pest and diseases, and very few diseases of plants survive in or on a seed. This predator and disease release is one reason why plants in a new country can do so well. Secondly, spreading your seeds reduces the risk from stochastic, or sudden and unpredictable, disasters. Thirdly, it reduces parent–child competition (more anthropomorphizing of botany). Finally, it increases the chance of finding a 'safe site'. Yet in reality, more than 80% of plants make no effort whatsoever to distribute their seeds, so having no adaptations for seed dispersal is no disadvantage until you have to migrate fast. If the climate is going to change fast, then fast migration may be selected for very positively in the next few decades and centuries.

Seeds are dispersal units, and by their ability to survive in soil seed banks they can be said to be dispersed in time. It is difficult to find any generalizations about which species of plant and which types of seeds you will find in a soil seed bank, partly due to the difficulty of finding the seeds. Generally, though, small seeds persist longer in the soil as they are less attractive to predators and because they more easily fall into the cracks in the soil than large seeds. Soil can hold many seeds if they are small enough, the record being 488,708 per square metre. The record for a seed surviving in soil is held by a seed of the sacred lotus, *Nelumbo*

**11. The seeds of *Nelumbo nucifer* remain viable for over 1,000 years**



*nucifera*, which germinated after 1,288 years (give or take 250 years). In the UK, an experiment was initiated by W. J. Beal in 1879. Viable seeds are still being extracted from the soil more than 120 years after they were buried.

Seeds in the soil seed bank may not be dormant, they may just be in the wrong conditions for germination. In this state, the seed is said to be quiescent, and for buried seeds light is what is missing. In order to persist, the seed may be dry, but some seeds survive better imbibed as they are then better able to repair routine damage inflicted on membranes and DNA. However, the longer that a seed remains buried, the greater the chance of death from pathogens, predators, or premature germination followed by grazing. The ability to persist appears to be a trait belonging to a species rather than individuals within a species. Persistence is also related to the potential risk of totally failing to get at least one plant into the next generation. This means that plants in stable communities like woodland show less persistence in the soil's seed bank than seeds in Mediterranean-type habitats.

The difficult concept of dormancy has already been mentioned and needs to be properly understood. A seed does not germinate in a packet of seeds because the minimum requirements for germination are absent. These seeds are quiescent. This prevents the seed from germinating if the conditions are wrong at a moment in time. Dormancy is a state of the seed and not the environment. If a seed does not germinate within one month of being sown, then it is either dead or dormant. If it is dormant, then that dormancy has to be broken for the seed to germinate. The role of dormancy is to time germination when the *seedling* has the maximum chance of survival. Dormancy is less important in habitats that have uniform and predictable conditions, as these conditions alone will prevent seeds from germinating. Most species in tropical moist woodland do not produce dormant seeds.



There are three basic types of dormancy. Morphological dormancy is when the embryo in the seed is immature when shed, such as in orchid seeds. Physical dormancy is when a hard seed coat prevents the uptake of water and keeps the embryo dry, such as seeds in the pea family. Physiological dormancy is when some chemical change is required, such as in those seeds of the rose family that require a period of chilling. The first two are non-reversible while the latter is, thereby permitting some flexibility. Morphological and physical dormancy never combine; morphological and physiological often combine to give morphophysiological dormancy (MPD); and physical and physiological rarely combine.

Dormancy can trap a plant in a particular habitat, making it particularly vulnerable to changes in climate should they occur. However, the conditions experienced by the parent plant can influence dormancy, and so plants of one species can demonstrate different dormancy states depending where they grow. For example, trees of *Pinus brutia* that grow on Crete show no dormancy and germinate in the autumn when the rains come, while seeds from trees in northern Greece require a period of chilling to prevent germination until the winter is over and the spring conditions have arrived.

A commonly held view is that physical dormancy imposed by a hard seed is broken down by a period of abrasion or attack by a microbe. There is no evidence to support this view. The breaking of this type of dormancy is much more controlled, with a variety of different structures providing a temperature-operated one-way valve that permits the ingress of water. This strategy has evolved in several different groups of plants independently. Another strategy that has evolved several times is parasitism, and seeds of parasitic species perform the very neat trick of perceiving when their host plant is nearby.

The imbibing of water is the first stage of germination. This is followed by a rapid increase in respiration and the mobilization of food reserves. The embryo starts to grow, and when the young root emerges, the seed has germinated and there is no going back. Germination can be stimulated by a number of factors. Fluctuation of temperatures between night and day can be a means by which a seed perceives a gap in the canopy above the seed. Likewise, the ability to perceive the quality and quantity of light falling on a seed will enable the timing of germination to be coordinated with a specific time of year, a particular depth of burial, or the degree of shade. Light sensitivity is imposed by the seed coat, and it is worth remembering that the seed coat is provided by the maternal plant. This is one way in which the conditions experienced by the parent plant can affect both the dormancy and the germination requirements of the seed.

Availability of water will be a critical requirement for germination. Not all seeds are dry and tolerant of desiccation. At least 7.4% of all species that produce seeds are recalcitrant, meaning that they will not tolerate being dried out. Soil nutrient levels are important for the germination of some seeds. High levels of nitrogen are needed by weed species. A flush of nitrogen often follows the creation of a gap in the canopy, so light, temperature, and nutrient status of the soil can be interconnected. A widespread mechanism for breaking dormancy and promoting germination is smoke, and this is even found in species that do not currently live in fire-dominated habitats.

It is often stated that the leaf litter from dominant plants can inhibit the germination of seeds. Although the idea of suppressing one's offspring is attractive, this form of allelopathy has never been demonstrated in a natural system. Leaf litter can provide a mechanical barrier to germination and may keep light-requiring seeds in the dark. The effect of leaf litter on germination and seedling establishment is a classic biological puzzle because, while litter can provide a good seed bed for germination, it can also be a habitat for small, grazing animals such as slugs and snails.



So germination is affected by temperature (actual and relative), water availability, nutrient levels (especially nitrogen), fire, and the presence of a soil seed bank. It is suspected that all of these will change with a change in the climate. The implications of climate change on seed behaviour could be profound.

Even when the seed has germinated, it is still in peril. Studies have shown that more than 95% of seedlings may be eaten or otherwise killed. This is on top of all the seeds that were eaten or fell in inappropriate places. These losses still may not be show-stoppers. The absence of a suitable gap may be the biggest single reason for the failure of a seed to grow into a mature plant. John Harper introduced the concept of *safe sites* where all the conditions are present for germination and successful establishment of the seedling. These sites may be very small, and perhaps just 1% of seeds ever find a safe site. The soil must be correct, the mycorrhizal fungus must be present and will form its association within seven to ten years, and light levels must be appropriate.

Seed biology may appear to be imprecise. We know far more about seeds today than we did twenty years ago, but there is still an absence of any rules. Is this because the production of seeds is such an evolutionarily risky strategy? The embryo is a vulnerable stage in any life cycle, but to release your embryo with the added attraction to predators of a food supply seems like odd behaviour for any parent, and yet it works against all the odds. It has been wisely suggested by Fenner and Thompson that there is 'a regeneration lottery that results in the maintenance of species diversity by default'. This lottery has resulted in more than 400,000 species of plants. How can this diversity be organized to aid our understanding of biology? Chapter 5 has an answer.

# Chapter 5
# Making sense of plant diversity

The naming of plants is the oldest profession. If you go back to the Book of Genesis, chapter 2, verse 19, God brought the products of his Creation to Adam and suggested that He had done His bit and that it was Adam's turn to name the results. Irrespective of your take on the verisimilitude of this historical account, it does show that the author of the Bible recognized the need for plants to have names if he was going to tell his story. Adam did the job very quickly because it was just three verses later that God made him a plaything for an idle moment. Adam was able to do the job quickly because he was starting with a clean slate and there was only him.

Today, the situation is far more complicated because plants have names given them by many indigenous populations without reference to anyone else. So ox-eye daisies in Buckinghamshire are called moon daisies in Oxfordshire on the other side of the Chiltern Hills. In Germany, there are 108 names for the white water lily. The word 'lily' is itself widely applied to plants such as water lilies, arum lilies, lily of the valley, and Madonna lilies. Only the latter is a true lily. There is an advantage to these vernacular names. They are in the local language and therefore more easily remembered. This is very important if the name is deadly nightshade. This name is, however, meaningless in France, where the same plant will have a different name. The problem that we have is that in a world where

**12. Ox-eye daisies have many different English names depending on where they live**

people travel around and talk to each other, there must be a universally accepted set of names. Linnaeus, the great global cataloguer of plant names, stated that 'without permanent names there can be no permanence of knowledge'.

Thanks to Linnaeus (and others), we do have permanent names and they are in the form of a Latin (or Greek) binomial. The great thing about Latin and ancient Greek is that they are apolitical languages so no one can complain on nationalistic grounds. They are also no longer evolving as languages. If you know some Latin or Greek, then you can start to appreciate the descriptive nature of plant names. *Calliandra haematocephala* is a plant in the pea family whose flowers are crowded together in a hemisphere. Each flower is superficially made up of nothing but stamens and so the inflorescence looks like a powder-puff, hence its common name. The Latin name means 'beautiful male parts', which is as good a name as powder-puff, but it does make *Calliandra* an inappropriate name for a girl, and yet .... One does feel some sympathy with the Chinese because so many of their plants are named after foreigners. *Mahonia fortunei*, for example, is named after a Pennsylvanian nursery man and a Scottish plant hunter.

**13. *Calliandra haematocephala* with flowers that appear to be made up of many male stamens, hence its Latin name meaning 'beautiful male parts'**

There are now rules that dictate how a plant should be named legitimately. Firstly, you check that is does not already have a name. When you are sure that it is a new plant species, you give it a name in Latin and describe it in Latin, though in 2011 it was decided by the International Botanical Congress that this latter condition should no longer be mandatory. Secondly, you make a dried pressed specimen and place this in a herbarium, where dried and pressed plant specimens are stored in perpetuity. Finally, when you publish the name and description of the plant in a peer-reviewed journal, PhD thesis, or similar, you state which herbarium holds the specimen so that botanists can return to see the actual plant that you named for at least 350 years. While there are rules, there is insufficient vetting of the names that are published, and so for every species of flowering plant there is on average two or three legitimate names. These are referred to as synonyms, with the most widely used name usually being the accepted name.

If you talk to botanists from around the world, you discover that many naming systems exist and that they are never random allocations of names, like children's Christian names. In China, for example, they have named all their magnolias in one group known as *xin ji hua*. In New Zealand, *Cordyline fruticosa* is known in Maori as Ti Plant and the varieties are known as Ti this, Ti that, and Ti the other. Binomial nomenclature is a common feature of many vernacular naming systems. As people named the plants that they saw around them, they intuitively placed similar species in groups. We see this in the book credited as being the beginning of modern botany – *Enquiry into Plants* by Theophrastus, published about 2,300 years ago.

Theophrastus records his observations of plants, and in the process puts them into groups on the basis of what they look like. For example, he describes a bunch of plants that have flowers all emanating from a single point, with much divided leaves that clasp the stem at the base of their stalk. These are all umbellifers or members of the celery family (the Apiaceae). This book was still

**14. The flower head of a hogweed plant. This belongs to a group of plants known as the umbellifers, and this group has been recognized since the 3rd century BC**



being published as a textbook in 1644, and translations and facsimiles are still available. The study of botany continued through the Greek and Roman empires, and in the middle 1st century AD. Dioscorides published *De Materia Medica*. This work remained in use until the 17th century. It described the plants used medically by Dioscorides, but the plants were grouped by their morphological characters as well as medical properties. (We shall see later that these two often go together.)

With the Renaissance in Europe and the exponential growth in travel came a renewed interest in plants for their own sake rather than just for their utilitarian value. Many herbals were produced in the 15th century and onwards, and these books are very good examples of a utilitarian classification. We still see these informal groups. In our gardens, we group plants on the basis of the conditions that they need (rock garden, water garden, woodland gardens, and so on) or on their colour (the red border at Hidcote Manor, or the White Garden at Sissinghurst Castle). We have vegetable gardens, herb gardens, and fruit orchards. In other societies, we see groups of edible plants, fibre plants, poisonous plants, and so on. These classifications are obviously helpful and will never die out, but they are not all-inclusive and they are not unique, so one plant may appear several times.

The early botanists, such as Swiss polymath Conrad Gesner (1516–65), set out to describe all of the natural world they saw around them. Gesner was a meticulous worker, and the annotated drawings and woodcuts that have survived show a terrific attention to detail. What set Gesner apart from other botanists of that time was that he used every character that he could observe rather than splitting plants into groups on one character alone. This obsession with finding one essential feature was a direct link back to Aristotelian philosophy, and it remained pre-eminent in plant classification until John Locke (1632–1704) introduced the ideas of observations being the route to knowledge and that we are all born knowing nothing. One person who was particularly influenced by Locke was John Ray (1627–1705).

Taxonomy is the science of placing living things in groups (*taxis* is Greek for order or arrangement), and John Ray is the father of modern taxonomy and the second greatest natural historian that England has produced. He was equally comfortable working with plants, animals, and rocks. He was severely troubled by fossils of animals that were no longer living – he referred to them as 'games of nature'. Yet despite his strong religious convictions, Ray laid down the ground rules that have underpinned all taxonomy since.

Ray was the first to define what he meant by a species. This may not seem startling and worthy of note today, because we use the word 'species' in everyday communication and so assume that the

**15. Middleton Hall in Staffordshire, where John Ray worked after leaving Cambridge**



word has always had a definition and meaning, yet it did not in Ray's time, and in fact even today there are still more than twenty definitions of the word. Species are the building blocks of taxonomy and the currency of biology, so a definition is critically important. Ray's belief was that:

> no surer criterion for determining species has occurred to me than the distinguishing features that perpetuate themselves in propagation from seed. Thus, no matter what variations occur in the individuals or the species, if they spring from the seed of one and the same plant, they are accidental variations and not such as to distinguish a species...; one species never springs from the seed of another nor vice versa.

So a species was a group of similar individuals that can freely interbreed to produce offspring that resemble their parents. This biological definition is still accepted by many and quoted in many exams. There are distinct problems with it when you try to apply it to orchids or to dandelions *inter alia*, so a simpler, workable definition is that 'a species is a group of individuals that share a unique set of characters that can be reproduced'.

So Ray recognized that if you wish to maintain the natural variation within a species it should be propagated by seed. This was truly prophetic, as this was 200 years before Mendel's work on genetics (and thus variation) was published and 300 years before the Millennium Seed Bank Project was established at Royal Botanical Gardens Kew to conserve 25% of the world's plant species in the form of seeds. However, Ray's biggest contribution to taxonomy arose from his assertion (following Locke's principles and the practice of Gesner) that if you wanted to arrive at a natural classification, you must use *every* character that you observe and/or measure. You should not ignore anything without good reason. His observations were limited by the resolution of microscopy at that time, yet his work represents a turning point in plant classification. (It should be noted that a *natural* classification to Ray and his contemporaries was one that reflected the *Creator's* plan.)

He dismissed the first division of plants into herbs and trees. He stated that all flowering plants belonged exclusively to one of two large groups that he termed the monocotyledonous plants (monocots) and the dicotyledonous plants (dicots). The monocots he described as those plants in which the seeds have one cotyledon, the floral parts are in multiples of three, the leaves have veins that are parallel, the stems have discrete bundles of vascular tissues scattered through the stem, and the roots are adventitious. On the other hand, dicots he described as those plants in which the seeds have two cotyledons, the floral parts are in multiples of two, four, or five, the leaves have a central vein with lateral and sublateral veins leading to a net-like arrangement, the stems have a ring of vascular tissues around the outer edge of the stem, and the roots are permanent with a persistent tap root and lateral roots. The monocot/dicot split

**16. A palm stem showing the bundles of vascular tissue typical of monocots**

persisted in plant classification until 1998, and it has to be said that the monocots as defined by Ray still exist. As a further result of Ray using every character that he could see, he also began to put plants into family groups that we still use. The borage family, for example, was one group Ray described, though he did not use the term 'family'.

Ray was by no means the only taxonomist working at this time. He regularly corresponded with Joseph Pitton de Tournefort (1656–1708) at the Jardin des Plantes in Paris. As a taxonomist, Tournefort was a trees and herbs man, but by looking at flower structures he did a great deal of tidying up at the next level up from species in the taxonomic hierarchy, namely the genus (plural genera). This set the scene for Carl Linnaeus (1707–78) to introduce universal Latin binomials, one for each species. During field work, Linnaeus found the long, detailed, descriptive polynomials very cumbersome. He proposed keeping Tournefort's genera but to reduce the rest of the name to a nickname that was hopefully helpful in identifying the plant or remembering it. In 1753, he published *Species Plantarum* which marked the start of legitimate names. He listed every species of which he was aware with one Latin binomial and all the Latin polynomials that had been applied to that plant since Theophrastus and Dioscorides.

It is interesting to note that in one of his four autobiographies, Linnaeus stated that he did not think that *Species Plantarum*, published in 1753, would be his legacy. For most scientists, this would have been enough of a legacy for one man, but Linnaeus was convinced that with his sexual system of classification described in *Systema Naturae*, published in 1735, he had hit upon the perfect natural classification for plants. He grouped them entirely on the number of male and female parts they possessed. His writings were very colourful, and perhaps he set out to shock as much as to enlighten. His sexual system of classification was never universally accepted, though it was useful if you were trying to identify a plant. It did, however, show that the division of plants into trees and herbs was dead and buried.

Linnaeus came up with the idea of classifying plants in this way when he was still in his twenties. He was trying to classify plants from the top down. This is when the characters that you are going to use are determined before you have looked at all the species in front of you. It is very common to find 21st-century undergraduates doing the same thing. John Ray, on the other hand, put all his plants into species with very full descriptions, and when you do this the genera and families almost select themselves; bottom up is better.

The next step forward in the development of our current system of classification was not taken by Linnaeus but by Antoine Laurent de Jussieu, who took the genera of Linnaeus, Tournefort, and others, and placed them in families. The account of this, published in 1789, marks the start of legitimate family names.

As the 19th century began, the plant kingdom had been grouped into species, species into genera, genera into families, families into orders, orders into classes, classes into phyla, and finally phyla into the kingdoms. This hierarchy of taxonomic ranks was a clear and simple way of organizing an ever-increasing amount of data. As plant hunters went about their work bringing plants back to Kew, Edinburgh, and the like, these new species could be inserted into the system. Each species had its own unique position and was given a group at each rank. This made identifying plants easier, either by asking a series of questions, such as it is a monocot or dicot, in order to narrow down the options, or by recognizing the family to which the plant belonged. The botanical family in particular is an exceptionally useful rank, as many of them, such as the orchid, daisy, pea, and celery families, are very easily recognizable.



However, there was revolution brewing as various people began to have doubts about the fixity of species. This was the belief that the world's species were fixed according to the Creator's plan and that they could not change. This went back to Ray's worries about fossils and beyond. Linnaeus had discovered that many plants displayed characters that were a mixture of two others. He often gave these the specific epithet of *hybridus* (for example, *Trifolium hybridus* that still grows on the route taken by Linnaeus when botanizing with his students). How could hybrids arise when all species had been created on Day 1 by God? Linnaeus very neatly got round this problem by suggesting that God created genera and Nature created the species.

As the son of a clergyman, perhaps Linnaeus knew when to keep his head down. However, people like Jean-Baptiste Lamarck (1744–1829) at the Jardin des Plantes were beginning to think openly that species changed in response to their environment. Lamarck is unfairly mocked by some, but Darwin and others give him the credit he deserves. He was one of the people who bravely kicked the evolutionary wasp nest and softened up the scientific world for the theory of evolution so brilliantly encapsulated in the first edition of Charles Darwin's *Origin of Species*.

This book should be compulsory reading for all first-year biology students, and no one should be allowed to begin a degree course until they have read it properly (with their i-pod switched off). The reason for this seemingly draconian idea is that, like Ray, Darwin was a natural historian, equally at home with plants as animals and rocks. His style of writing is still a classic of science interpretation. *Provoke, relate, reveal* is the mantra of conservation education, and that is just what Darwin does. By the time you get to Chapter 13 of *On the Origin of Species*, you are completely drawn into Darwin's line of thought.

Chapter 13 is relevant here, as it is the chapter in which Darwin considers classification. He believed that if his idea of descent with modification was correct, then this would go a long way to explaining the fact that we had classified biology into this hierarchical system of nested ranks. The only diagram in *On the Origin of Species* is a branching tree – the era of phylogenies had begun. Darwin believed that species could be placed in genera because they shared, at some point in the past, a common ancestor – *commonality of descent* – and that the further you had to go back to find that common ancestor, the further up the hierarchy of ranks you went – *propinquity of descent*. The system of classification did not prove evolution had happened, but the system of evolution proposed by Darwin explained why we could classify as we did. At a stroke, the meaning of *natural classification* had changed. No longer were taxonomists trying to reveal the Creator's plan; now they were replaying evolutionary history. This in itself was one of the great objections to Darwin's theory, because whereas the Creator was believed to have had a plan, Evolution has no plan, and that frightened people.

As the 19th century rolled on, the number of plants in herbaria such as at Kew grew, and the classification was adjusted and refined. Between 1862 and 1883, George Bentham and Sir Joseph Hooker published their classification of plants. In the flowering plants, they recognized monocots and dicots. Within the dicots, the first divisions were into three big groups: plants whose flowers had petals that are free from each other, plants where the petals are fused into a tube, and those where there appear to be no petals. Within the monocots, the divisions were based on an eclectic range of characters such as whether the flowers were colourful or brown and membranous, whether the seeds were tiny, and so on. Bentham and Hooker's system was very useful for identification purposes, but they made no attempt to reveal evolution despite Hooker being a major correspondent with Darwin. Perhaps it was just too controversial.



The publication of *On the Origin of Species* opened the door for the creation of phylogenies. German polymath Ernst Haekel (1834–1919) is often given the credit for coining the term 'phylogeny'. A phylogeny is an evolutionary classification and is drawn as a branching tree with each end branch representing a group. It is hoped that one day we shall be able to create a phylogeny for all flowering plants and there will be about 350,000 terminal branches, each one representing a species. Haekel's trees were based on the morphology that he and others could see with increasingly sophisticated microscopes, so essentially they were using similar evidence to that available to Ray and his colleagues.

In the 20th century, a new possibility arose. With the discovery of the role of chromosomes and then the structure of DNA in 1952, followed by the deciphering of the genetic code, people began to wonder if this was the essence about which Aristotle had written. Would it be possible to find a stretch of DNA that is unique to every member of a species? Could we use the difference in the sequence of bases in a stretch of DNA taken from two different organisms to determine their propinquity of descent? These are in fact two different things. The first is identification, the second is classification. Before either of these questions could be answered, another technical innovation rocked the taxonomic boat, and this was the invention of the electron microscope that revealed a box full of new characters to be considered.

One of the most beautiful plant structures revealed by electron microscopists is the patterning on the outside of pollen grains. This is the sporopollenin that has featured so much through this book. The photographs taken with electron microscopes revealed not only patterns that can be unique to a particular species, but also that pollen grains have a number of apertures through which the pollen tube grows after the stigmatic surface has given it water. The number of apertures varies from one to several, but it is usually one or three. This was a hugely significant character because it helped to solve a problem that had become increasingly awkward.

Along with the addition of structures revealed by electron microscopes and the addition of DNA sequences (also known as molecular data), there had been one very major shift in the rules. This was and is the principle of monophyly. A monophyletic group is one that contains all of the descendants of a common ancestor. To put it another way, all the taxa in a monophyletic group share a common ancestor, and the group must contain all of the descendants of that common ancestor. Monophyly can be applied at every rank. So all living organisms form a monophyletic group because life has evolved once. Plants (as defined in this book) are a monophyletic group, as are land plants, seed plants, flowering plants, monocots, orchids, and so on. None of the new evidence of characters has shaken the monophyletic status of the monocots as described by John Ray over 300 years ago. However, the dicots have not fared so well.

When Darwin began to think about how particular lineages might be related, he came up against a problem. Although seed plants (gymnosperms and angiosperms) appeared to be a monophyletic group, that is to say that seeds have evolved just once, what did the common ancestor look like, and what was the origin of the angiosperms? He described this to Hooker as 'an abominable mystery', and mystery it remains. One way of trying to reveal the truth is to find out what the first flowering plants looked like. Before the molecular data and electron micrographs became available, botanists were beginning to think that magnolias represented an early group because the oldest fossils of flowering plants looked very similar to modern magnolias, and because magnolias are pollinated by beetles, which were present before bees and were around when these fossil flowers were living. In addition to the magnolias, the waterlilies, peppers, bays, and several other groups were being examined. All of them were dicots. The pollen evidence then threw a large spanner in the works. It turned out that all these groups which were suspected of being in some way odd all had pollen with just one opening, like the monocots. All the dicots which appeared to be in the clear had pollen with three apertures.



When the molecular data was thrown into the mix with all the macro-morphology and micro-morphology, it appeared that the flowering plants are made up of two very big monophyletic groups, the monocots and the true dicots (officially the eudicots), and lots of little groups. This seemingly heretical suggestion was first aired in 1993, and it was considered so radical that half of the contributing scientists withdrew their names from the paper. By the time the second version was published in 1998, no one was in any doubt that the monocots and eudicots were here to stay, but the others needed to be sorted out. It is now thought that the 352,000 species of flowering plants should be grouped into 13,500 genera in 462 families in 41 orders (or thereabouts!).

The 1998 paper was a landmark in plant taxonomy. Firstly, it had a huge author list because it was a multinational collaboration, nominally led by Mark Chase from the Royal Botanic Gardens Kew. However, rather than being cited as Chase *et al.*, the paper's authorship is APG – the acronym for the Angiosperm Phylogeny Group. Secondly, the paper was rejected by both *Nature* and *Science*, leaving the Missouri Botanic Garden to publish it in their *Annals*. The biggest paper in plant taxonomy for 150 years got more coverage in the UK press than in the elite scientific journals, clearly illustrating the depth to which taxonomy had sunk in the eyes of the scientific establishment. Thirdly, the APG said they did not know where to place certain groups. This was unheard of. All previous classifications were complete, with every taxa assigned to a place at every rank. Since APG I, there have been APG II (2002) and APG III (2009). The detail is becoming clearer. The lowest branch on the flowering plant phylogeny is *Amborella* from New Caledonia. Being the lowest branch means that this species had a common ancestor with all other species of flowering plants longer than any other group of flowering plant. They are often referred to as a relic group. Next is the waterlilies and a few allies; next the Austrobaileyales order, a group of families including most famously star anise, which for several years was the origin of tamiflu, a drug that was going to treat the pandemic of swine flu in 2009. After that, the picture is murky. There appears to be a large group consisting of the magnolias, black peppers, bay, and avocados, but we do not know how this relates to the monocots and dicots or a couple of oddities, *Ceratophyllum* and *Chloranthus*. Only time will tell, but as yet Darwin's abominable mystery is still a mystery.

After all this work to classify the world's plants, one might ask 'so what?'. Is this just a job-creation scheme for vegetation trainspotters? Is a subject with an acknowledged subjective and philosophical component really a science? The latter is easily answered. Taxonomy is trying to make order out of random evolution. Species are constantly changing. There is no fundamental difference between a species and a variety of that species (Darwin, 1859). Taxonomy is therefore like nailing

**17. *Magnolia* flowers appear to have existed for at least 115 million years**

jelly to the wall – messy, but worthwhile for the following reasons.

Firstly, a phylogenetic classification enables us to make objective decisions in conservation where inadequate resources have to be allocated to the places of greatest need. Take, for example, two similar areas of vegetation each containing 8,000 species but you can only conserve one area. If one contained plants from 100 families in 20 orders, and the other plants from 50 families in 10 orders, then the former probably contains more of the evolutionary tree than the latter and so is more valuable.

Secondly, as indicated earlier, some parts of the evolutionary tree are richer in medically efficacious materials than others. The potato family (the Solanaceae) has given us atropine and hyoscyamine, among others. When the power of galanthamine to treat Alzheimer's disease was discovered, it was only known from snowdrops. It was also known that snowdrops could not grow fast enough to supply the huge demand. An alternative supplier was quickly found in the Amaryllidaceae, in the shape of daffodils. Likewise, although taxol for the treatment of ovarian cancer was originally extracted from the Pacific yew (*Taxus brevifolia*), it is not a sustainable supply, and an alternative supply of the raw material has been found and commercially extracted from English yew (*Taxus baccata*). In the past year, harringtonine has been licensed for the treatment of leukaemia. Harringtonine was first found in *Cephalotaxus*, a close relative of yew trees.

But drugs are not the only things we get from plants, so what have plants ever done for us?

Chapter 6

# What have plants ever done for us?

It has been said that there are three broad reasons for conserving plants: the 3Ss or the 3Ps or the 4Es. These are sensible/selfish/spiritual or pragmatic/practical/philosophical or ecological/economic/ethical/aesthetic (well almost). There is little doubt that plants (and biology in general) improve the quality of our lives. They are beautiful, intriguing, and their cultivation is regarded by many people to be therapeutic. The role of plants in art and the art of landscape design cannot be ignored, but these are personal aspects of plants that fall under the spiritual/philosophical/aesthetic banner.

This leaves two broad groups of reasons why plants are important. On the one hand, there are the big ecological services that are normally presented under four headings: provisioning (e.g. food or fresh water), regulating (e.g. carbon sequestration), supporting (e.g. providing pollinators), and finally cultural (e.g. hiking). To a degree, these services are not species specific, so it can be argued that it does not matter which tree species is growing and absorbing carbon so long as a tree is growing. It has been calculated that the value of theses ecosystem services is approximately $33,000,000 million per annum. This is a strange, sterile calculation because if you had $33,000,000 million in you wallet, where would you go to buy these ecosystem services?

Furthermore, the global gross national product for the planet each year is just $18,000,000 million, so even if it was for sale, we could not afford it. If something is not available for purchase, is it priceless or valueless? Perhaps this accounts for the cavalier manner in which biology has been treated in the past few centuries.

The other reasons for looking after our biological heritage are species specific. These are those commodities and resources that we extract or otherwise derive directly from the herbage. Included here are plants like the English yew, which supplies us with Bacattin III used in the synthesis of taxotere for the treatment of breast cancer.

Asking what have plants done for us may have become a question that needed to be answered only in recent decades as some of us become increasingly removed from the plants that we exploit. On one simple level, plants have done, and will continue to do, everything for us because of their (as yet) unique ability to use the Sun's energy to make an awe-inspiring range of chemicals, some of which all the Queen's chemists cannot synthesize.

On another, very biological, level, plants create the habitat that we are currently occupying with at least 1.5 million other species. Yet one of the things that sets *Homo sapiens* apart from other species is its ability to increase the carrying capacity of its habitat. It has been suggested that in the long term, sustainable agriculture can only support 1,500 million people, though in the medium term it will be required to support a predicted maximum 9,500 million by 2050. Without agriculture, the planet will support approximately 30 million hunter-gatherers. This is a sobering (or just silly) figure since there is nowhere for the other 6,470 million people to go. In order to feed the current population, we are cultivating one-quarter of the total land area, and we are currently consuming, in one way or another, 40% of all photosynthetic activity. It is of great concern that if we have to expand the area under cultivation, then many other species are going to be evicted from their current habitats.

The origin of agriculture is by no means clear. The traditional view that persisted until the end of the 20th century was that about 8,000 years ago (YBP) grasses were domesticated in the region between the River Tigris and the River Euphrates, henceforth known as the Fertile Crescent. It is believed that as many as eight crops were grown here: einkorn wheat, emmer wheat, barley, lentils, peas, flax, bitter vetch, and chickpeas, with beans appearing later. At about the same time, rice was domesticated in China and potatoes bred in South America. After that time, more independent centres of domestication emerged, including the breeding of maize in Central America from teosinte. What could have stimulated this simultaneous, but uncoordinated, adoption

**18. Sweetcorn is now one of the three major crops**

of an agrarian way of life? The climatic aberration known as the Younger Dryas Event was traditionally given the credit for being the catalyst. This simple view is now being challenged.

It is now thought that the end of the last ice age, c. 14,500 YBP, was marked, in the Middle East and probably elsewhere, by an increase in temperature closer to present values. The temperature then promptly returned to ice-age values about c. 13,000 YBP. This is known as the Bølling-Allerød Interval. This was followed by the cold but dry Younger Dryas Event, c. 12,900 to c. 11,600 YBP, at the end of which the temperature rose to far above that of the ice age and the glaciers finally retreated. Following the end of the Younger Dryas, the climate became unfamiliarly constant from year to year, and so humans stopped moving around in groups of between 15 and 50 and they started to live in settlements that were increasingly made of stone. These are known as the Natufian people and they lived in what is now Israel, Jordan, Syria, and the Lebanon. Olives, pistachios, wheat, and barley began growing in this area as the temperature rose. These plants were pre-adapted to these new conditions.

There are now about 60 sites that were populated by the Natufians, but there is very little direct evidence about which plants were being used. One of these sites is at Abu Hureyra in the Euphrates Valley in Syria. There are two sets of remains here from before and after the Younger Dryas Event. It is thought by some that it was the cold, dry weather that forced the people in these areas to cultivate grains, which had become much rarer in the wild. Nine plump rye seeds have been found at this site. Some researchers say this is hardly firm foundations for an entire theory, not least because rye has not been found elsewhere and does not make another appearance for a few millennia. It may be that the 50% increase in carbon dioxide levels at the end of the Younger Dryas Event helped by making agriculture photosynthetically supportable.

There is some agreement about how *Homo sapiens* brought plants under his control. We know that hunter-gatherers ate fruits, roots, seeds, and nuts from wild plants. Included here are grains from grasses. In 2009, evidence was presented that appears to show that grains, and in particular sorghum, were being ground up using stone tools in Mozambique 105,000 YBP. This evidence is supported by the presence of amyloplasts in the tools. Amyloplasts are the organelles where plants make and store starch. The pattern of the deposition of the starch and the size of the amyloplasts is often species specific. Furthermore, cultivated varieties often have larger amyloplasts. The amyloplasts of wild chillies are 0.006 millimetres long, while cultivated varieties are 0.02 millimetres long. Amyloplasts are really useful because they are resilient to decay and are found not only in early kitchen utensils but also in sediments. They have now been used to date the domestication of squashes, manioc, and chilli peppers in America.



On the south-west shore of Galilee is the Ohalo II site, which was inhabited 23,000 YBP. More than 90,000 plant remnants have been found here from *inter alia* acorns, pistachios, wild olive, and lots of wheat and barley, but nothing that looks like a cultivated variety (cultivar), and there is no evidence for the grinding up of cereals.

It is important to this story to consider what is meant by cultivation and domestication. The former is simply the deliberate growing or protection of a wild plant. Growing begins with a propagule such as a seed or cutting, while protection is where a particular plant is preserved and nurtured because it has something that is wanted. It is often difficult to prove cultivation of wild plants. Domestication, on the other hand, is visible. There are many definitions for domestication, but they all include an element of permanent physical and genetic alteration to improve the plant and make it more suitable for the needs of humans.

These improvements are known as domestication syndromes. Some of the characters are obviously better for the farmer. These include compact growth, making the plant more resilient, simultaneous ripening allowing one harvest, loss of seasonal flowering and/or germination permitting sowing at various times of year, and increase in the size of the grain with a thinner seed coat for increased palatability and easier processing such as grinding and milling. Enhanced flavour and nutritional value are characters that the consumer would like to see, and when agriculture began the farmer and consumer were the same people (as they still are in many parts of the world).

For some crops, there are very specific obstacles to overcome during domestication. In figs, this is the requirement for the flowers to be pollinated by a specific species of wasp before the 'fruit' that we eat can develop properly. The flowers are enclosed in the syconium (the fleshy part of the fig that we eat and which is, strictly speaking, not a true fruit). The figs that we buy in greengrocers are a mutant variety that will develop its fleshy syconium without the need for the pollination of its flowers. These parthenocarpic varieties are ancient. Remains of figs excavated at Gilgall, a neolithic village in Jordan, show extraordinary detail of one of these ancient parthenocarpic fruits that has been accurately dated at 11,400 years old.

One character that can frustrate farmers is the propensity for plants to drop their seeds when they are ripe. The farmer wants the plant to hang on to its seeds and fruit until he is ready to harvest them. This means that seeds from domesticated varieties do not have a smooth abscission wound but a jagged scar or tear. This is a very useful trait for archaeobotanists because it can be found on the remains of very old seeds. Such seeds have been found in the remains of the Nevali Cori settlement in Turkey. Seeds found here were harvested 10,500 YBP, and they have a jagged abscission layer indicating that they did not fall naturally but were torn off during the harvest. This is the earliest direct evidence for the domestication of plants.

However, it should not be assumed that the transition from collecting wild grains to growing fields of selected and bred varieties happened cleanly one year. There is good evidence, also from Nevali Cori, that wild and domesticated plants were grown side by side, perhaps because the farmer could not prevent some of the wild seeds from falling before harvest thus enabling them to persist from year to year in the soil seed bank. It is now widely accepted that domestication was preceded by many years of cultivation and so a possible scenario for the transition from hunting and gathering to farming may have involved four stages. The first involved the hunter-gatherers moving around in small groups, their movements dictated by availability of food and by the weather, which could be very inconsistent. These wanderers would inhabit caves where possible, and in these is found debris from their everyday lives. More than 1,000 fibres of flax have been found in caves in the Caucasus Mountains. These sites were active 36,000 YBP. These fibres appear to have been dyed, even though dying flax is very difficult.

It is known that *Homo neanderthalensis* and subsequently *Homo sapiens* had learned that fire could be used to great advantage. Selective burning of small areas provoked the regrowth of soft, palatable vegetation. Lignified tissue is almost impossible to digest, but cellulose poses less of a problem to some grazing animals. This means that the regenerating burned area would attract animals that could be picked off easily by men with spears waiting in the unburned bushes.

The second stage in the adoption of an agrarian way of live would have involved a combination of hunting, gathering, and cultivation of wild plants. In this way, the skill of farming would have been learned and understood.

Having learned how to grow plants, the early farmers were able to turn their attention to the selection of better plants by looking for the domestication syndromes described earlier. Thus, the third stage of the process involved a decreasing reliance on hunting and gathering and more dependence on the produce of fields. These fields would have contained an increasing proportion of domesticated varieties of both plants and animals. Evidence from the Ohalo II site indicates that 10,000 YBP, just 10% of cultivated plants exhibited domestication syndrome. This proportion had risen to 36% by 8,500 YBP and to 64% by 7,500 YBP. It is also becoming clear that different crops were domesticated at different rates, that those rates were not constant but punctuated with bursts of activity, and that different crops were altered in different ways.

It is now believed that although the Chinese began consuming rice 12,000 YBP, it was in fact millet that they began to domesticate first, around 10,000 YBP. In the Chengdu Plain in western China, millet was the staple diet 4,000 YBP and was used to make flour, porridge, and beer. Evidence from China shows that the domestication of rice was a slow process. Around 24,000 pieces of plant have been examined from just one site, including 2,600 rice spikelets. While rice domestication may have started 10,000 YBP, by 6,900 YBP only 8% of cultivated plants were rice and just 27.4% of this was domesticated. By 6,600 YBP, rice accounted for 24% of all cultivated plants but only 38.8% was domesticated. As cultivation spread, so did the extent to which the environment was altered to accommodate fields. In the lower Yangtze region, alder woodland was cleared c. 7,700 YBP and the water levels managed for 150 years by the use of bunds until a disastrous flood c. 7,500 YBP.

The final stage in the evolution of *Homo sapiens* into an agriculture-dependent species involved an increase in the area of land under cultivation and a spreading of this way of life. It is now believed that although squash, peanuts, and manioc were growing in the Andes 10,000 YBP, they were originally domesticated in the lowland tropical forests. The evidence for this comes from the study of the genomes of the wild and cultivated forms. There is also evidence of earthworks in western Amazonia that is the proposed centre for the domestication of not only squash, peanuts, and manioc, but also chilli, beans, rubber, tobacco, and cocoa.

It is believed that maize was being grown in the Andes between 7,000 and 5,000 YBP. However, we do not know how different these plants were from the original teosinte from which modern maize has been bred. While modern wheat and barley both appear to have been bred in more than one place, with more than one original hybrid or selection, modern maize all stems from one domestication event, at some time between 9,000 and 6,000 YBP. The transformation is eye-watering. For example, the maximum number of kernels on teosinte is about 12, whereas a cob of corn will now contain up to 500 or more as a result of selecting plants with more and more kernels and growing and crossing only those plants.



It is now believed that there are more than 24 regions where cultivation began independently. In 13 of these areas, grains were the major crops. As a rule of thumb, increasing seed size and non-shattering seed heads were the first two domestication syndromes to be worked on. For pepo squashes, it was not only larger seeds that were selected but also thicker stems and thinner rinds.

Following the emergence of these two dozen centres of domestication, agriculture began to spread all over the world as people moved about. Sometimes they took their plants with them. For example, the African gourd (*Lagenaria siceraria*) has been found in 10,000-year-old settlements in America. It is believed that it was taken there by Palaeoindians perhaps via the coast of Asia rather than across the Atlantic. There is good evidence that the early farmers in Europe were migrants who taught the indigenous population how to farm. It is not clear yet whence they came but by bringing cultivated varieties of plants with them, and thus taking the cultivars away for their parent species, they were inadvertently speeding up the process of domestication by preventing genetic dilution of their selections by back-crossing with the parent stock. In effect, they were separating the gene pools.

While food plants were being domesticated, other natural resources were being extracted from plants. Rubber and fibres have already been mentioned, but trees were exploited for their timber and their leaves in the construction of dwellings and the heating of those homes. However, another major area of use was in medicines.

It is difficult to the say when and how early hominids began to appreciate the healing power of plants. Again, it is debris in caves that gives us some clues. In caves in northern Iraq, pollen grains are found in large numbers, and it is assumed these are from the plants that were exploited by the cave's inhabitants, be they *Homo neanderthalensis*, *Homo sapiens*, or the recently discovered hybrids. About 90% of the pollen grains have been identified as being from plant species still used in this part of the world for their medicinal properties. How these properties were discovered is a subject for speculation, but there are several possibilities.

Serendipity cannot be ruled out, as simple chance observation; certainly, this was how people would have discovered which plants were toxic. Observing animals may have made people inquisitive about plants. Echinacea, willows, and tea tree are eaten by mongoose, horses, and koala respectively. All have proven beneficial effects. Various strange ideas have made a small contribution to our knowledge. In pole position here must be the Doctrine of Signatures, based on the idea that if a part of a plant resembles a part of the body, then the plant should be used to treat ailments of that part of the body. Only *Aristolochia* used in encouraging

**19. Birthwort has been used for centuries to induce childbirth**



childbirth, and *Podophyllum* used to treat testicular cancer, can be claimed as pseudo-proof for this charmingly dangerous idea.

Traditional Chinese medicine is probably the most widely used form of herbal medicine. The oldest printed herbal is the *Pen Tsao* printed in China 4,000 years ago. Many of the infusions and tinctures recommended in this work are still used in China today and still work. One of the problems for science is discovering the active ingredients in these preparations because they often contain more than one plant, and so the active ingredient may be a combination of several ingredients working together. A very extensive programme is currently underway in China that aims to uncover the active ingredients in all 6,000 species that are used in traditional Chinese medicine.

The Chinese have used *Ephedra* for more than three millennia in the control of bronchial asthma, and it is still a common ingredient in children's cough medicine. It is also used by European anaesthetists to prevent the patient coughing during the operation, something that could be bad news. They also use it as the patient comes round as it mimics the effect of adrenalin, thereby making them feel re-invigorated and ready to go home. It should be noted that the International Olympic Committee and others regard ephedrine and its derivatives as a performance-enhancing substance. The use of *Artemisia annua* in the treatment and prevention of malaria has been derived straight from Chinese herbal medicine, though the mode of action is still poorly understood. For many years, extract of *Indigofera* has been used to treat 'bad blood', and it is now used in the treatment of leukaemia in the West.

The use of *Ginkgo biloba* in China to treat circulatory problems, especially to the brain, remains to be proved, but the author's mother found it was terrific for improving circulation to her hands and feet when nothing prescribed by the GP had worked. However, her heart is being kept going by the extraordinary ability of digitoxin to regulate and strengthen heart beat. The original use of *Digitalis purpurea*, and more recently *Digitalis lanata* digoxin and then digitoxin, was first proved by a Dr Withering working in the village of Edgebaston in the 19th century, when Birmingham was smaller and cricket had not been created. He had come across the idea from a Shropshire Romany tincture from 20 herbs that was claimed to be efficacious against a wide range of ailments. By process of elimination, the foxglove was found to be the hero of the hour.



With the advent of the scientific method, more rigorous testing was carried out on the claims made about the therapeutic value of plant extracts. The problem was finding willing subjects upon whom ideas could be tested. In the 16th century, an early form of the parole board in prisons enabled prisoners to be released early if they allowed physicians to observe the effects of various treatments on the inmate. Early release was guaranteed – but

**20. The flowers of English foxgloves**

often only in a coffin. The toxic properties of hellebores were discovered in this way.

Clinical trials are now more tightly controlled, and the search for new treatments more systematic. However, the value of local knowledge is still inestimable. The discovery of taxol is a direct result of the native North Americans sharing their traditional medicinal knowledge with the US National Cancer Institute. Having extracted the taxol from the bark of *Taxus brevifolia* and demonstrated its potential against ovarian cancer in mice, the research was taken up by a commercial pharmaceutical company. Their first problem was getting hold of sufficient taxol to carry out the research. It was not possible to synthesize the stuff, and so the only option was to harvest it from the bark of the trees. While this helped the company, and ultimately the patient, it did nothing for the trees. This was never going to be a sustainable supply of taxol.

In Europe, where researchers had no access to the trees of *Taxus brevifolia* (because they belong to the American State), an alternative production system had to be found. Eventually, the leaves of English yew (*Taxus bacatta*) was found to contain Bacattin III, which can be changed into medically efficacious taxol and subsequently taxotere for the treatment of breast cancer. In addition to showing that new, modern, clinically proven treatments are still being derived from plants, this story raises three other important issues. Firstly, we should look after all species. Even if we do not exploit them at present, we may in the future. Secondly, the sustainable extraction of biological resources is very important. There is no biological credit card. That being said, we are using up our reserves at an impressive rate. It has been calculated that in a single year, we release carbon dioxide from fossil fuels that took plants three million years to fix by photosynthesis. Thirdly, this raises the question of ownership. The US took the view that taxol is theirs and no-one else can exploit it. (In the end, it transpired that they have rights to the ownership of *Taxus brevifolia* but not the patent on taxol.) These three issues are the three headings under which the Convention on Biological Diversity derives its framework.

Living within our biological means is an important issue that will not go away. Photosynthesis is the only way of providing us with our daily food and so much else. It has been possible to increase the output per acre for many of our staple foods. The most recent green revolution was led by Norman Bourlaug and others. In 1970, Bourlaug was awarded the Nobel Peace Prize in recognition of the impact that he had had on crop yields in those parts of the world where malnutrition was a way of life. Between 1966 and 1968, the yield of wheat in India rose from 12 million tonnes to 17 million tonnes. A plant breeding programme for rice resulted in a similar increase in yields. Sadly, the results in Asia were never transferred to African agriculture, perhaps as a result of the lack of infrastructure such as roads, irrigation, and seed production. Bourlaug's mantra of 'impact on farmers' fields, not learned publications, is the measure by which we will judge the value of our work' is not quite in line with current science research funding in the UK.

It is estimated that between 1960 and 2000, the proportion of the world's population who felt hungry for at least a part of the year fell from 60% to 14%, though this is still almost 1,000 million people. The next green revolution will have to increase yields further. This may be possible if waste and loss to pests and diseases can be reduced. Growing drought-tolerant crops such as sorghum might be another option. One exciting possibility is to alter the photosynthetic machinery in rice to make it more like maize. (If you are motivated to learn more about this, rice is C3 and maize C4, which means that rice fixes carbon dioxide into a molecule with three carbon atoms whereas maize fixes carbon dioxide into a molecule with four carbon atoms.) This would require some clever modification of the genetics of the rice. Such a modification appears to have happened 45 times in the course of evolution, but this is different, and some people are frightened of further genetic modification of our food plants.

The risks of genetically modified crops fall into two broad categories: human and animal protection and ecological protection. There are already procedures to ensure that food sold in the UK is safe to eat. Ecological protection is more difficult but can be tested and assessed. The three major risks may be summarized thus. Will the genetically modified variety escape from its field? Will it hybridize with native species? Will it cross-pollinate non-genetically modified crops of the same crop plant? If the answer to all of these is no, then perhaps the risks are acceptable. Ultimately, it will be the lack of an alternative that will drive people to embrace genetically modified crops and the food that results.

The green revolution of the 1960s and beyond relied on high inputs of water and fertilizer. We know that phosphorus supplies may peter out in the middle of the 21st century, closely followed by the supplies of inorganic nitrogen. Fresh water supplies are not infinite. This is a particularly relevant problem in the drive to produce bioethanol and biodiesel for our vehicles. It has been calculated that while oil extraction and refining may use up to 190 litres of water per kilowatt hour, and nuclear power up to 950 litres of water per kilowatt hour, the production of corn ethanal requires up to 8,670,000 litres of water per kilowatt hour, and soybean biodiesel production up to 27,900,000 litres of water per kilowatt hour. It appears that biofuels will make us all very thirsty.

# Chapter 7
# Looking after the plants that support us

On 5 June 1992 in Rio de Janeiro, at the first Earth Summit, the Convention on Biological Diversity (CBD) was opened. To date, the vast majority of nation states have signed and ratified the convention that aims to conserve the components of biodiversity, to ensure its sustainable exploitation and to facilitate the equitable sharing of the benefits of this exploitation. As a result of this landmark agreement, there was a huge increase in conservation activity *sensu lato*. Each country drew up an action plan and took responsibility for the implementation of the CBD in their region. That this happened is supported by the fact that 90% of money raised for conservation is spent in that money's country of origin. An example of this is the Millennium Seed Bank Project at the Royal Botanic Gardens Kew that had as its first target the collection and safe storage of almost all the around 1,500 species of plant native to the UK. In addition to the increase in practical activities, there was an explosion of scientific papers mentioning biodiversity in their title.

By the end of the 20th century, it became apparent to many botanists, especially those working in botanic gardens around the world, that although there was a great deal of activity, there was no way of knowing if the aspirations of the CBD were being met. Following a meeting in April 2000, the Gran Canaria declaration was published calling for a Global Strategy for Plant Conservation

(GSPC) and suggesting more than a dozen targets to be hit by 2010. This idea was discussed by the countries that had ratified the CBD, and in September 2002 the GSPC was adopted, with 16 targets to be hit by 2010. This was perhaps a surprise, as never before had a target-orientated strategy been proposed for a major group of organisms. The rest of the conservation community is very interested to see if this approach works because it may be a model for all conservation work in the future. By 2010, it was clear that some of the targets had been hit, some were going to be missed but not by much, and that a couple were going to be missed badly. As a result, GSPC 2 was drawn up during 2010. This was clearly based on the successes and failures of GSPC 1; the progress of plant conservation was much better informed in 2010 than it was ten years earlier.

The GSPC is a wonderful framework, or list, of what we need to achieve if we are going to realize the aspirations of the CBD in relation to plants. The 16 targets cover all areas of activity related to plant conservation. It enables different organizations to make a contribution commensurate with their resources and missions; it is acknowledged that neither one organization nor one person is going to save all of the world's plant species from extinction. In order both to understand the scale of the task to halt and then reverse the decline in plant species and to appreciate that this is not an insurmountable problem, it is helpful to go through the targets of the GSPC.

Target one is to draw up an online list of all the known plant species in the world. This task is being coordinated by the Royal Botanic Gardens Kew and the Missouri Botanical Garden which between them have the most comprehensive herbarium collections in the world. The problem of compiling this list is that there are two ways of cataloguing species. One is by monographs that survey all the species in a genus (or part of a genus, if the genus is large). This work is generally carried out by one person who takes a global view of the diversity within the species. The other type of catalogue is the national inventory found in a flora. While national pride is often a potent driving force behind conservation work, it is well known that local botanists will often seek to inflate the number of species in their country by raising to the rank of species small variations that are just the normal range of diversity found within a species. This innocent chauvinism means that you cannot simply add up all the species in floras. Nor can you just make a list of all the legitimately described species, because most species have been named more than once and so there are synonyms to sort out.

The work to compile the world checklist, as a first stage in the creation of a world flora, has been organized by botanical family and this has thrown up some anomalies. There appears to be no pattern to the areas of ignorance. Large families are no more difficult than small families. Families that have a long history of cultivation, for either economic reasons or aesthetic pleasure, are no more likely to be well understood than those that are currently of botanical interest only. This work is therefore showing where there is a lack of expertise – this is referred to as the 'taxonomic impediment' in Europe and as the 'taxonomic abyss' in Australia!



Having put together the list of species, we need to know which are struggling to survive and this is target two. There is a long tradition of countries and regions compiling 'red lists' of threatened species under the aegis of the International Union for the Conservation of Nature (IUCN). These have been compiled on a national basis for many countries, but bizarrely these suffer from the same inflation seen in floras. It appears that long national red lists are seen as a way of levering more money out of funding bodies. National lists also lead to species that are very common in one country but rare next door appearing to be threatened; false positives are common.

There is a global list that includes an assessment of 47,677 species across all taxonomic groups. This list currently (2009) shows that

17,291 species, or 36%, are threatened with extinction in the next 50 years. The statistics for plants are rather worse, with 71% of the 12,151 species assessed being either extinct, extinct in the wild, critically endangered, endangered, or vulnerable, with just 12% being categorized as least concern. These figures reveal one of the problems connected with data-gathering, that of definitions. The IUCN categories are clearly defined and are the gold standard, but they take a great deal of time to compile, time we may not have.

It has become apparent that an heuristic approach is required in the first instance. Such a system has been devised, the RAMAS Red List, that reduces to just three categories: likely threatened, likely not threatened, and likely data-deficient. Or to put it another way: threatened, OK, or don't know. This allows workers to input data quickly, without the need for lengthy calculations and comprehensive details. In addition to providing a more broad-based approximation of the problem, this will show where the data deficiencies are and thus where more resources are needed. It is now expected that all new monographs will include an assessment of the threatened status of all the species described. For some genera, such as magnolias and maples, genera-based assessments have been carried out with full monographic revisions. Other estimates of the number of threatened species exist, and generally the numbers are between 22% and 47%. It is reasonable to assume, therefore, that the general 36% figure is not unreasonable, and so if there are 352,828 species of flowering plants, then 127,018 are threatened with extinction in the next 50 years.

It is already clear that there is a great deal of work being done, and more is required if just targets one and two are to be hit. Sound scientific research in all aspects of conservation biology is required, and the execution and dissemination of this research is target 3. While peer-reviewed journals are always going to have their place, it is equally true that there is great deal of work going unreported and yet which would be very helpful to conservation workers, especially those in the field of developing countries. Since 2002, a number of web-based resources have emerged where evidence-based conservation can be posted. A recurring experience of conservation projects is that each one is slightly different. Each species recovery project is different and each habitat restoration project is different because the assemblage of species is different from site to site. While the CBD celebrates and protects biological diversity, that same diversity makes generalizations difficult and makes rules loose.

So the first three targets are about understanding and documentation without which the next seven targets would be unattainable. Target four is to protect at least 15% of the world's main ecological regions. It is currently estimated that 11.5% of the land area of the world is under some form of protection. In China, this figure is 14.7%. The advantage of conserving ecological regions is that you conserve the ecosystem services described in the previous chapter. Furthermore, it is often claimed that if you have a healthy, stable, diverse community of plants then the animals, fungi, and other organisms will be there too. It is for this reason that the diversity of plant species is often used as a surrogate measure for all of biodiversity. This is not unreasonable, as most ecosystems are described by the plants that dominate them. It is furthermore claimed that by protecting large areas, you are protecting not only the species that you have recorded, you are also looking after the unknowns. This is particularly true for insects, of which perhaps only 10% have been described. It is not as important for plants, as it is believed that 90% of all plant species have been named at least once.

While the protection of ecosystem services is undeniably important, it is equally true that plant diversity in general, and individual species distributions in particular, are not evenly spread throughout the world. This problem is addressed by target five, whereby 75% of the important plant areas are to be protected. The objective here is to try to avoid gaps in the coverage of plants at all latitudes and types of habitat. It is well known that, generally speaking, the number of plant species per unit of area falls as you get further from the equator, but this does not mean that the temperate regions are less important and valuable than the tropical regions. It has also been found that deciding where to place your protected areas can require some very complicated analysis of data. A study in Madagascar attempted to optimize the allocation of protected areas for 2,138 species of plants and animals. They were only permitted to protect 10% of the island, but each group of plant and animal required a slightly different 10%. In the Philippines, another study showed that while many plants were adequately protected in nature reserves, the threatened palm trees grew mostly outside these areas.

While protecting areas of 'natural' vegetation will always be seen as a good idea, it should not be forgotten that 25% of the land on

**21. Palmer's Leys is a field in Oxfordshire that has been restored to wild flower meadow in just three years. Since 1950, 96% of this type of meadow in the UK has been ploughed up**

Earth is under some type of production regime. This does not have to be an intensive agricultural system with high inputs of fertilizers, pesticides, and perhaps water. It can be a hill farm in Wales or a cork oak woodland in Iberia. The latter is a very good example of a commercially supportable production system that has an associated flora that puts it in the top 20 regions of the world for plant diversity. Target six aims to have 75% of production lands managed with the conservation of plant diversity as one of the aims. This is more easily done for woodland than for a farmer's field, and yet stone curlews are often seen nesting in intensive sugarbeet fields in East Anglia. Worldwide, as much as 60% of forestry land has the conservation of the biodiversity written into the management goals for the area.

So these three targets are concerned with large-scale, vegetation-based conservation. However, much of conservation is at the level of species, not least because people can see that they are making a difference if they champion and protect a species. There is a risk that species-based conservation is not coordinated and the species chosen for conservation are not the most deserving but perhaps the most iconic. It is, for example, easier to fund orchid conservation than the protection of a species of moss. That being said, a great deal of conservation work is carried out at the level of species.



For many years, the ultimate goal was to carry out all species conservation *in situ*, and if the species was declining in its habitat then you would grow some more and reintroduce it. This failed far more than it succeeded because, unless the threat or reason for the decline is removed, then the new plants will go the same way as their predecessors. The only threat that could be removed easily was over-collecting by humans. This led to people considering *ex situ* conservation whereby plants were grown outside their habitat, in, for example, a botanic garden or arboretum. As a result, conservation became polarized into two camps, and targets seven and eight of the GSPC reflect this division. Target seven is that 75% of threatened species are conserved *in situ*, while in target eight, 75% of threatened species are conserved *ex situ*, with 10% of these projects in the country of origin of the species concerned.

There is no reason why a declining species should not be conserved in a managed nature reserve. The advantage of this method is that all the other relationships that the plant has with fungi, pollinators, dispersal agents, and other plants are maintained. Furthermore, it is generally believed that more genetic diversity is preserved in *in situ* populations and that the species will therefore be able to evolve as it adapts to changes in conditions. This may all be true, but plants may be far more tolerant of changing pollinators than we previously thought, few plants use a dispersing agent, and the number of plants needed to preserve the genetic diversity of a population of plants is far fewer than you might imagine. Somewhere between 50 and 500 unrelated plants are all you need to preserve 95% of the genetic diversity of a population. The major disadvantage of *in situ* conservation is the security of the site. In the past, this was seen as protection from anthropogenic development or habitat transformation in general. It is now seen that there are two more threats to sites that may be more difficult to control. One is invasion by non-native species, more of which later. The other is climate change.

That the climate is changing cannot be seriously disputed. Whether *Homo sapiens* is behind it and can therefore control it is not important here. The world's plants have experienced climate change before, most recently during the shenanigans at the end of the last ice age. Plants coped presumably by a combination of migration, adaptation, and using hitherto unexploited traits. The relative contributions of these strategies probably varied from species to species and from habitat to habitat. If migration is important, and we know that many plants did migrate long distances during the last three million years, then the encroachment of the urban and agricultural landscapes into natural habitats may have scuppered any chance of migration. The idea of changing the scale of the human landscape at this stage is impractical, but the policy of conserving plants in static nature reserves is equally problematic. Corridors linking them is an idea often proposed, as is landscape conservation whereby the positioning of reserves is coordinated.

All of the advantages of *in situ* conservation can be flipped into the disadvantages of *ex situ* programmes. However, there is a half-way-house known as a species-recovery programme where the goal is to establish a self-propagating population of a species somewhere. In order to achieve this goal, you need to be able to supply everything that the plant requires. You need to understand the habitat requirements, pollination biology, seed storage needs, and you need someone to provide the ongoing, endless monitoring to ensure that the plant survives. This strategy has been championed in Western Australia, where so much cutting-edge conservation happens. Plant species have been brought back from the edge of extinction by species-recovery programmes such as this.



*Ex situ* conservation has always been regarded as the poor relation to *in situ* conservation, but maybe its time has come in the form of seed banks. While the Millennium Seed Bank Project's (MSBP) first target was the UK flora, it had a secondary target of banking 25,000 species from around the world. That target has been reached and a new more ambitious goal defined. In a world where the future climate is such a big known unknown, seed banks like the MSBP look like a very good idea. All of the criticisms such as genetic erosion, loss of viability, vulnerability to stochastic events like war can be addressed by proper collecting protocols, research, and duplicate collections. It is true to say that much of what we have learnt in the past two decades about seeds is as a result of the rise of seed banks. It may be that in the future seeds stored in seed banks will be used to facilitate assisted migration of plants from

**22. Seedlings of the endangered *Encephalartos ferox* being cultivated *ex situ***



one nature reserve to another. It is true that not all seeds appreciate being dried and frozen. Perhaps as many as 30% of species have these so-called recalcitrant seeds, and for these alternatives need to be found.

Seed banks are not a new idea. They began life as gene banks for cultivated varieties of major crop plants, and worldwide there are a number of banks for major crops such as wheat, rice, and maize as well as vegetables. Many of these crops are annuals and so very suited to storage and to the goal of target nine of the GSPC that 70% of the genetic diversity of our crop plants is preserved and protected. Some crops are not propagated by seed, so for crops such as potatoes and fruit trees, field gene banks are the alternative. In many countries, amateur gardeners are keeping the older varieties alive. These may have unique traits that we shall need in the future to impart disease resistance or tolerance of drought.

Mention has already been made of the problems posed by invasive non-native species. This is one of the five major causes of the losses of biological diversity known acronymically as HIPPO: Habitat transformation, Invasive species, Pollution, Population growth, and Over-exploitation. Darwin's theory of evolution through natural selection of organisms that are better suited to their circumstances has often been misrepresented by the sound-bite 'survival of the fittest' when it should be 'survival of the slightly better and the lucky ones'. Darwin observed 150 years ago that nowhere in the world was so well stocked, and the resources so well exploited, that there was no room for another species from abroad. He cited the Cape of South Africa which perhaps has a higher density of species than anywhere else on the planet.

Some people are worried that the fact that the organisms best adapted to grow in South Africa actually evolved in Australia and Europe and vice versa. There are many possible reasons why a non-native plant can outcompete the natives. One is the concept of predator release. This refers to the possibility that plants leave behind the herbivores, pests, and diseases that control them in their native lands and in their new abode they are free from attack. Whatever the reasons, there is a 1 in 1,000 probability that a plant brought to Country B from Country A will become an invasive species and thus cause irreversible damage to the plant communities in Country B. This may not seem short odds, but 70,000 different plants are currently on sale in UK nurseries. These are fully functioning genomes that are potentially far more harmful to other plants than a maize plant that has been genetically modified to resist a herbicide.

Target ten of GSPC 1 was to draw up control measures for the 100 most damaging non-native species. This target has been met, and in GSPC 2 it was modified to be more specific about the control of non-natives in Important Plant Areas in all countries, and more specifically control of further invasions. A problem encountered when controlling non-natives species is predicting which will become invasive. There are reasons why species do not invade; the soil is the wrong pH, the winters too cold, the summers too dry, pollinators fail, dispersal fails, and so on. Unfortunately, though, there is no blueprint for a potential invader; it is only possible to be wise after the event – it is very much like economics in that respect. The only control is prevention. Borders should be sealed to non-native species, as already happens in the USA, Australia, and New Zealand.

The CBD is very clear that we are being greedy when it comes to our consumption of the products of photosynthesis. The Convention on the International Trade in Endangered Species (CITES) should already prevent the exploitation of declining species, but target eleven reiterates that no species of plant should be threatened by trade. Sometimes it is difficult to know which plant is being traded. How can you identify a species when it has been made into a set of window blinds or a herbal remedy? The answer may be close at hand. The practice of DNA barcoding of species has been successful in animals. The idea is that you find a stretch of DNA that varies very little between members of a species, but which varies far more between species. All you have to do then is extract some DNA from the selected region of the genome, sequence it, compare it to the library of this sequence in every known species, and you will either get a name for your sample or be told that you have found a new species. This sounds like science fiction, but it is nearly reality and the prototypes for hand-held machines to do this in the field are being tested. The next step will be to create the library of sequences.

While target eleven addresses the protection of species that are already endangered, the original target twelve aimed at preventing other species being added to the list by trying to ensure that at least 30% of all plant-based products are derived from sustainably harvested suppliers. The revised target twelve is that all wild harvest, plant-based products must be sourced sustainably. Anyone who purchases anything can contribute to the attainment of this target. There are a number of schemes, such as organic food and the FSC (Forestry Stewardship Council) scheme, that enable you to reduce your impact on the world's biological resources. As with many aspects of biology, this area of the GSPC is neither black nor white. The controversy over the production of palm oil in regions of the world inhabited by orang-utans is one such problem. Taking away someone's ability to generate an income is serious action and should not be undertaken lightly.

Hand in hand with the loss of biological resources often goes local, indigenous knowledge. We have a range of phrases that covers this type of information and rarely is it complimentary. Old wives' tales or folklore are just two. It is common to find that this specialist ethnobotanical knowledge is not written down and so is vulnerable when communities are displaced or attracted by the promises of Western knowledge and the Western way of life. Had it not been for the work of ethnobotanists, it may never have been found that the sap of *Strophanthus*, long used as an arrow-tip poison, contained strophanthine that is widely used as a treatment for heart disorders. Target thirteen hopes to stem the tide of ignorance resulting from the careless loss of knowledge, but, with the previous target, this will not be achieved by GSPC 1.

At this point, one comes up against the perennial argument about the interface between poverty and the conservation of biology. Views range from those who think that poverty alleviation is always going to be more important than the existence of a plant species, to those who would be happy to see starvation if that is what the future of plants requires. The reality is, as ever, somewhere in between, and a number of economists are now realizing that economic prosperity is often underpinned by healthy biodiversity. However, to state that poor people are responsible for the destruction of biology and rich people are all caring would be very unfair. The regions of the world where plant species are most threatened can be linked to neither prosperity nor poverty.

**23. *Strophanthus*, from which strophanthine is extracted to treat heart disease**



The last three targets are too open-ended for us to be sure if a difference has been made, but there are good examples of work that is contributing. Target fourteen urges that the need to conserve our plant heritage is included in education programmes wherever possible. Plants do not have their own voices and so they need an evangelist to speak for them. Botanic gardens, arboreta, and wildlife trusts are just three types of organizations that provide public education programmes that aim to highlight the need to conserve plants and their habitats. Another way to realize the aspirations of the target is to have the conservation of plants written into national curricula for biology courses, and perhaps economics and geography *inter alia*.

The distribution of plants across the world is very uneven and so is the distribution of those people working to conserve these plants on the targets of the GSPC. Target fifteen aims to redress the balance and to ensure that there is a sufficient number of people trained to fulfil the targets and that these people are adequately resourced. It is too commonly stated that no one studies botany any more. If you base this statement on simply the *names* of university courses in the UK, then you may be correct. However, if you take the time to read syllabuses, you will quickly discover that courses in biology, environmental science, conservation, and geography often include all the elements of a 1960s and 1970s botany course. The word 'botany' has a veneer of Victorian parsonages and Edwardian ladies that repels potential students in a way that the newer courses do not. In some countries, however, botany is in the ascendancy. Two such countries are Brazil and Ethiopia. Ten years ago, the project to produce a flora of Ethiopia was staffed entirely by foreigners. Now the work is being carried out entirely by Ethiopians.

It is true to say that projects such as the Millennium Seed Bank have been very effective at creating partnerships and building

**24. Training the next generation of field botanists is seen as a priority for many botanic gardens**

capacity for plant conservation in many countries. It is vital that the people who are involved in all aspects of plant conservation and the GSPC are working in a coordinated manner. International, regional, national, and local networks are required and are being formed in line with the proposals in the final target, number sixteen. A newly formed network will link people working in the conservation of the plants of oceanic islands. These biological gems often have very high levels of endemism and thus unique floras. They often share common threats. Invasions by non-native species are especially common and damaging on islands. These networks will bring together state organizations, NGOs, international agencies, and local people, as well as amateur botanists and natural historians who are a vital component of many successful national strategies.

No one person or organization is going to solve the problem of declining species on their own. Now that we have the GSPC, we see more clearly where plant conservation has been successful and where the next round of work must be concentrated. It has become a blueprint for other areas of conservation.

We are now in a position to say with confidence that there is no technical reason why a plant species cannot be saved from extinction. In some countries, almost every target of the GSPC will be hit. One such country is the UK, where 400 years of botany and a tradition of field work and botanical recording shows that if there is a sufficient commitment from the indigenous population, then there is a good future for plants and for the humans who will forever depend on these plants for everything.

# Further reading

David Beerling, *The Emerald Planet: How Plants Changed Earth's History* (Oxford University Press, 2007)

Eric Chivian and Aaron Bernstein, *Sustaining Life: How Human Health Depends on Biodiversity* (Oxford University Press, 2008)

Charles Darwin *The Origin of Species* (1859)

Jordan Goodman and Vivien Walsh, *The Story of Taxol: Nature and Politics in the Pursuit of an Anti-Cancer Drug* (Cambridge University Press, 2001)

Nicholas Harberd, *Seed to Seed: The Secret Life of Plants* (Bloomsbury, 2007)

Vernon Heywood (ed.), *Flowering Plant Families of the World* (Firefly Books, 2007)

Anna Lewington, *Plants for People* (Eden Project Books, 2003)

D. J. Mabberley, *Mabberley's Plant-Book* (Cambridge University Press, 2008)

Oliver Morton, *Eating the Sun: How Plants Power the Planet* (Fourth Estate, 2007)

Denis J. Murphy, *People, Plants and Genes* (Oxford University Press, 2007)

Michael Proctor, Peter Yeo, and Andrew Lack, *The Natural History of Pollination* (Timber Press, 1996)

Michael G. Simpson, *Plant Systematics* (Elsevier Academic Press, 2006)

Alison M. Smith et al., *Plant Biology* (Garland Science, 2010)

Richard Southwood, *The Story of Life* (Oxford University Press, 2003)

William J. Sutherland and David A. Hill, *Managing Habitats for Conservation* (Cambridge University Press, 1995)
Joel L. Swerdlow, *Nature's Medicine: Plants That Heal* (National Geographic, 2000)
Thomas N. Taylor, Edith L. Taylor, and Michael Krings, *Paleobotany: The Biology and Evolution of Fossil Plants* (Academic Press, 2009)

# “牛津通识读本”已出书目

古典哲学的趣味
人生的意义
文学理论入门
大众经济学
历史之源
设计，无处不在
生活中的心理学
政治的历史与边界
哲学的思与惑
资本主义
美国总统制
海德格尔
我们时代的伦理学
卡夫卡是谁
考古学的过去与未来
天文学简史
社会学的意识
康德
尼采
亚里士多德的世界
西方艺术新论
全球化面面观
简明逻辑学
法哲学：价值与事实
政治哲学与幸福根基
选择理论
后殖民主义与世界格局
福柯
缤纷的语言学
达达和超现实主义
佛学概论
维特根斯坦与哲学
科学哲学
印度哲学祛魅
克尔凯郭尔
科学革命
广告
数学
叔本华
笛卡尔
基督教神学
犹太人与犹太教
现代日本
罗兰·巴特
马基雅维里
全球经济史
进化
性存在
量子理论
牛顿新传
国际移民
哈贝马斯
医学伦理
黑格尔
地球
记忆
法律
中国文学
托克维尔
休谟
分子
法国大革命
丝绸之路
民族主义
科幻作品
罗素
美国政党与选举
美国最高法院
纪录片
大萧条与罗斯福新政
领导力
无神论
罗马共和国
美国国会
民主
英格兰文学
现代主义
网络
自闭症
德里达
浪漫主义

批判理论
电影
俄罗斯文学
古典文学
大数据
洛克
幸福
免疫系统
银行学
德国文学
戏剧
腐败
医事法
癌症
植物
法语文学
微观经济学
湖泊
儿童心理学
时装
现代拉丁美洲文学
卢梭
隐私
电影音乐
抑郁症
传染病